Communications in Computer and Information Science 1050

Commenced Publication in 2007
Founding and Former Series Editors:
Phoebe Chen, Alfredo Cuzzocrea, Xiaoyong Du, Orhun Kara, Ting Liu,
Krishna M. Sivalingam, Dominik Ślęzak, Takashi Washio, and Xiaokang Yang

More information about this series at http://www.springer.com/series/7899

Marcelo Naiouf · Franco Chichizola ·
Enzo Rucci (Eds.)

Cloud Computing and Big Data

7th Conference, JCC&BD 2019
La Plata, Buenos Aires, Argentina, June 24–28, 2019
Revised Selected Papers

 Springer

Editors
Marcelo Naiouf (iD)
III-LIDI, Facultad de Informatica
Universidad Nacional de La Plata
La Plata, Argentina

Franco Chichizola (iD)
III-LIDI, Facultad de Informatica
Universidad Nacional de La Plata
La Plata, Argentina

Enzo Rucci (iD)
III-LIDI, Facultad de Informatica
Universidad Nacional de La Plata
La Plata, Argentina

ISSN 1865-0929   ISSN 1865-0937 (electronic)
Communications in Computer and Information Science
ISBN 978-3-030-27712-3   ISBN 978-3-030-27713-0 (eBook)
https://doi.org/10.1007/978-3-030-27713-0

This Springer imprint is published by the registered company Springer Nature Switzerland AG
The registered company address is: Gewerbestrasse 11, 6330 Cham, Switzerland

Preface

Welcome to the proceedings collection of the 7th Conference on Cloud Computing & Big Data (JCC&BD 2019), held in La Plata, Argentina, during June 24–28, 2019. JCC&BD 2019 was organized by the School of Computer Science of the National University of La Plata (UNLP).

Since 2013, the Conference on Cloud Computing Conference & Big Data (JCC&BD) has been an annual meeting where ideas, projects, scientific results, and applications in the Cloud Computing & Big Data areas are exchanged and disseminated. The conference focuses on the topics that allow interaction between the academy, industry, and other interested parts.

JCC&BD2019 covered the following topics: cloud, edge, fog, accelerator, green and mobile computing; cloud infrastructure and virtualization; data analytics, data intelligence, and data visualization; machine and deep learning; and special topics related to emerging technologies. In addition, special activities were also carried out, including one plenary lecture, one discussion panel, and two post-graduate courses.

In this edition, the conference received 31 submissions. All the accepted papers were peer-reviewed by three referees (single-blind review) and evaluated on the basis of technical quality, relevance, significance, and clarity. According to the recommendations of the reviewers, 12 of them were selected for this book (39% acceptance rate). We hope readers will find these contributions useful and inspiring for their future research.

Special thanks to all the people who contributed to the conference's success: Program and Organizing Committees, reviewers, speakers, authors, and all conference attendees. Finally, we would like to thank Springer for their support in publishing this book.

June 2019

Marcelo Naiouf
Franco Chichizola
Enzo Rucci

Organization

General Chair

Armando De Giusti   Universidad Nacional de La Plata-CONICET, Argentina

Program Committee Chairs

Marcelo Naiouf   Universidad Nacional de La Plata, Argentina
Franco Chichizola   Universidad Nacional de La Plata, Argentina
Enzo Rucci   Universidad Nacional de La Plata, Argentina

Program Committee

María José Abásolo   Universidad Nacional de La Plata-CIC, Argentina
José Aguilar   Universidad de Los Andes, Venezuela
Jorge Ardenghi   Universidad Nacional del Sur, Argentina
Javier Balladini   Universidad Nacional del Comahue, Argentina
Oscar Bria   Universidad Nacional de La Plata-INVAP, Argentina
Silvia Castro   Universidad Nacional del Sur, Argentina
Laura De Giusti   Universidad Nacional de La Plata-CIC, Argentina
Mónica Denham   Universidad Nacional de Río Negro-CONICET, Argentina
Javier Diaz   Universidad Nacional de La Plata, Argentina
Ramón Doallo   Universidad da Coruña, Spain
Marcelo Errecalde   Universidad Nacional de San Luis, Argentina
Elsa Estevez   Universidad Nacional del Sur-CONICET, Argentina
Aurelio Fernandez Bariviera   Universitat Rovira i Virgili, Spain
Fernando Emmanuel Frati   Universidad Nacional de Chilecito, Argentina
Carlos Garcia Garino   Universidad Nacional de Cuyo, Argentina
Adriana Angélica Gaudiani   Universidad Nacional de General Sarmiento, Argentina
Graciela Verónica Gil Costa   Universidad Nacional de San Luis-CONICET, Argentina
Roberto Guerrero   Universidad Nacional de San Luis, Argentina
Waldo Hasperué   Universidad Nacional de La Plata-CIC, Argentina
Francisco Daniel Igual Peña   Universidad Complutense de Madrid, Spain
Laura Lanzarini   Universidad Nacional de La Plata, Argentina
Guillermo Leguizamón   Universidad Nacional de San Luis, Argentina
Emilio Luque Fadón   Universidad Autónoma de Barcelona, Spain

Mauricio Marín	Universidad de Santiago de Chile, Chile
Luis Marrone	Universidad Nacional de La Plata, Argentina
Katzalin Olcoz Herrero	Universidad Complutense de Madrid, Spain
José Angel Olivas Varela	Universidad de Castilla-La Mancha, Spain
Xoan Pardo	Universidad da Coruña, Spain
María Fabiana Piccoli	Universidad Nacional de San Luis, Argentina
Luis Piñuel	Universidad Complutense de Madrid, Spain
Adrian Pousa	Universidad Nacional de La Plata, Argentina
Marcela Printista	Universidad Nacional de San Luis, Argentina
Dolores Isabel Rexachs del Rosario	Universidad Autónoma de Barcelona, Spain
Nelson Rodríguez	Universidad Nacional de San Juan, Argentina
Juan Carlos Saez Alcaide	Universidad Complutense de Madrid, Spain
Victoria Sanz	Universidad Nacional de La Plata, Argentina
Remo Suppi	Universidad Autónoma de Barcelona, Spain
Francisco Tirado Fernández	Universidad Complutense de Madrid, Spain
Juan Touriño Dominguez	Universidad da Coruña, Spain
Gonzalo Zarza	Globant, Argentina

Sponsors

COMISIÓN DE INVESTIGACIONES CIENTÍFICAS
Ministerio de Ciencia, Tecnología e Innovación

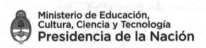
Ministerio de Educación, Cultura, Ciencia y Tecnología
Presidencia de la Nación

Sistema Nacional de Computación de Alto Desempeño

AGENCIA
Agencia Nacional de Promoción Científica y Tecnológica

RedUNCI
Red de Universidades Nacionales con Carreras de Informática

Contents

Cloud Computing and HPC

Detecting Time-Fragmented Cache Attacks Against AES Using
Performance Monitoring Counters.............................. 3
   Iván Prada, Francisco D. Igual, and Katzalin Olcoz

Hybrid Elastic ARM&Cloud HPC Collaborative Platform
for Generic Tasks.. 16
   David Petrocelli, Armando De Giusti, and Marcelo Naiouf

Benchmark Based on Application Signature to Analyze and Predict
Their Behavior.. 28
   Felipe Tirado, Alvaro Wong, Dolores Rexachs, and Emilio Luque

Evaluating Performance of Web Applications in (Cloud)
Virtualized Environments.................................. 41
   Fernando G. Tinetti and Christian Rodríguez

Intelligent Distributed System for Energy Efficient Control 51
   Martín Pi Puig, Juan Manuel Paniego, Santiago Medina,
   Sebastián Rodríguez Eguren, Leandro Libutti, Julieta Lanciotti,
   Joaquin De Antueno, Cesar Estrebou, Franco Chichizola,
   and Laura De Giusti

Heap-Based Algorithms to Accelerate Fingerprint Matching
on Parallel Platforms..................................... 61
   Ricardo J. Barrientos, Ruber Hernández-García, Kevin Ortega,
   Emilio Luque, and Daniel Peralta

Big Data and Data Intelligence

An Analysis of Local and Global Solutions to Address Big Data
Imbalanced Classification: A Case Study with SMOTE Preprocessing 75
   María José Basgall, Waldo Hasperué, Marcelo Naiouf,
   Alberto Fernández, and Francisco Herrera

Data Analytics for the Cryptocurrencies Behavior 86
   Eduardo Sánchez, Jose A. Olivas, and Francisco P. Romero

Measuring (in)variances in Convolutional Networks 98
   Facundo Quiroga, Jordina Torrents-Barrena, Laura Lanzarini,
   and Domenec Puig

Database NewSQL Performance Evaluation for Big Data
in the Public Cloud   110
    María Murazzo, Pablo Gómez, Nelson Rodríguez, and Diego Medel

Mobile Computing

A Study of Non-functional Requirements in Apps for Mobile Devices   125
    Leonardo Corbalán, Pablo Thomas, Lisandro Delía, Germán Cáseres,
    Juan Fernández Sosa, Fernando Tesone, and Patricia Pesado

Mobile and Wearable Computing in Patagonian Wilderness   137
    Samuel Almonacid, María R. Klagges, Pablo Navarro,
    Leonardo Morales, Bruno Pazos, Alexandra Contreras Puigbó,
    and Diego Firmenich

Author Index   155

Cloud Computing and HPC

Detecting Time-Fragmented Cache Attacks Against AES Using Performance Monitoring Counters

Iván Prada$^{(\boxtimes)}$, Francisco D. Igual, and Katzalin Olcoz

Departamento de Arquitectura de Computadores y Automática,
Universidad Complutense de Madrid, 28040 Madrid, Spain
{ivprada,figual,katzalin}@ucm.es

Abstract. Cache timing attacks use shared caches in multi-core processors as side channels to extract information from victim processes. These attacks are particularly dangerous in cloud infrastructures, in which the deployed countermeasures cause collateral effects in terms of performance loss and increase in energy consumption. We propose to monitor the victim process using an independent monitoring (detector) process, that continuously measures selected Performance Monitoring Counters (PMC) to detect the presence of an attack. Ad-hoc countermeasures can be applied only when such a risky situation arises. In our case, the victim process is the Advanced Encryption Standard (AES) encryption algorithm and the attack is performed by means of random encryption requests. We demonstrate that PMCs are a feasible tool to detect the attack and that sampling PMCs at high frequencies is worse than sampling at lower frequencies in terms of detection capabilities, particularly when the attack is fragmented in time to try to be hidden from detection.

Keywords: Cache attacks · Flush+reload · AES ·
Performance Monitoring Counters

1 Introduction

In January 2018, Horn [9] from Google Project Zero and a group of researchers led by Paul Kocher independently disclosed three vulnerabilities, named Spectre (variants 1 and 2) and Meltdown. They discovered that data cache timing could be used to extract information about memory contents using speculative execution. Since that moment, new variants of these transient execution attacks have been disclosed, such as Foreshadow or NetSpectre, to name just two of them [5].

These attacks exploit speculative and out-of-order execution in high performance microarchitectures together with the fact that in modern multi-core architectures some resources are shared across cores. Hence, a malicious process which is being executed in one core of the system can extract information from

© Springer Nature Switzerland AG 2019
M. Naiouf et al. (Eds.): JCC&BD 2019, CCIS 1050, pp. 3–15, 2019.
https://doi.org/10.1007/978-3-030-27713-0_1

a victim executed in a different core. The resource that is most commonly used as side-channel to extract information is the shared cache [2].

This problem is particularly important in cloud environments, where not only multiple users share a multi-core server but also multiple virtual machines can co-reside in the same core due to consolidation in order to save energy. Moreover, the use of simultaneous multithreading techniques, such as Intel's Hyperthreading technology, allow to leverage two or more logical cores per physical core, increasing the degree of resources shared between users.

There has been a proliferation of ad-hoc defenses, mainly microcode and software patches for the operating system and virtual machine monitor. Besides, Intel announced hardware mitigations in its Cascade Lake processors, trying to reduce performance loss due to the countermeasures for some of the attacks [11].

However, the impact of countermeasures on performance is still non negligible, and according to [5] varies from 0% to almost 75%. Thus, in most situations, security comes at the expense of lower performance and higher energy consumption (due to non-consolidating and disabling hyperthreading).

In this paper, we propose a new attack detection tool that is based on the deployment of a process running in the same core that the victim process it protects, and that detects situations in which an attack is being performed. Following this idea, countermeasures are only taken when the risk level justifies the cost.

The contribution of the paper is two-fold:

- We implement and describe the attack, and design and implement a detector for it based on Performance Monitoring Counters (PMC), evaluating its detection capabilities at different sampling frequencies and showing that high sampling frequencies (100 μs) are noisier than lower ones (10 to 100 ms).
- We show that splitting the attack into small pieces and distributing those pieces in time decreases detection capability in a different way for the different detection sampling frequencies. Only low frequencies (bigger than 10 ms) are still able to detect the time-fragmented attack.

The rest of the paper is structured as follows: Sect. 2 reviews the most relevant works in the field; Sect. 3 outlines the main concepts needed for the correct understanding of the attack and detection strategy. Then, the attack implemented, detection using PMC and the time-fragmented attack are presented in Sects. 4, 5 and 6, respectively. Finally, conclusions are presented in Sect. 7.

2   Related Work

Detailed surveys on microarchitectural timing attacks in general [2,8] and cache timing attacks in particular [12] can be found in the literature. Besides, Canella et al. [5] performs a systematic evaluation of transient execution attacks.

Time-driven attacks against the shared and inclusive Last Level Cache (LLC) are mainly based on Flush&Reload [15] and their variants. So,

Briongos et al. [4] extracts the key from the AES T-table based encryption algorithm using improvements over the original attack.

Recently, Performance Monitoring Counters have been used to detect the attack. Chiappeta et al. [6] monitor both victim and attacker, while CloudRadar [16] monitors all the virtual machines running in the system. CacheShield [3] only monitors the victim process to detect attacks on both AES and RSA algorithms. None of them considered trying to hide the attack by dividing it into small pieces distributed in time. Our approach is similar to CacheShield [3] in terms of functionality, but we perform a more detailed study of how the specific timing of the attack affects the detection capability.

3   Background Concepts

For a correct understanding of the attacks and techniques described hereafter, further details on two architectural concepts with direct impact on the attacks are required: *cache inclusion policies* and *memory de-duplication* as a specific case of shared memory. Then, the basics of the Flush&Reload attack are outlined.

3.1   Shared Caches and Inclusion Policies in Modern Multi-cores

Modern multi-core processors feature multi-level caches in which levels can be classified as *shared/private* across cores and hierarchies as *inclusive, non-inclusive* or *exclusive*, depending on whether the content of a cache level is present in lower cache levels. Of special interest for us is the combination of shared/inclusive cache levels, such as LLC caches in modern Intel multi-cores; in this scenario, a process executed on a specific core can produce side effects on independent processes executed on a different core. This phenomena can be exploited to perform cache-timing attacks. Supplementary techniques, such as Intel's Cache Allocation Technology (CAT [13]), can be leveraged to isolate specific LLC ways in order to boost performance (reducing contention), but also to mitigate the effects of potential attacks in this type of processors and situations.

3.2   Shared Memory and Memory De-duplication

Modern operating systems, such as Linux, make an intensive use of shared memory across processes to improve memory usage efficiency. Some situations (e.g. parent-child process hierarchies generated through *fork()*) are easily trackable, but sharing memory pages across independent processes requires ad-hoc sophisticated techniques. This is a very common scenario in multi-virtual machine (multi-VM) deployments sharing the same physical resources, for example.

Memory de-duplication is a specific technique of shared memory, designed to reduce the memory footprint in scenarios in which a hypervisor shares memory pages with the same contents across different virtual machines, but with impact also on non-virtualized environments comprising random non-related processes. In the Linux implementation (KSM, *Kernel Samepage Merging*), a kernel thread periodically checks every page in registered memory sections, and calculates a hash of its contents. This hash is then used to search other pages with identical contents. Upon success, pages are considered identical and merged, saving memory space. Processes that reference to the original pages are updated to point to the merged one. Only after a write operation from one of the VMs (or processes), sharing finishes and the corresponding page is copied by COW (*Copy-on-Write*).

3.3   Flush&Reload

The *Flush&Reload* attack was first introduced in [15] and it has been used as baseline by later works such as [10], among others. It takes advantage of the combination of inclusive shared caches and memory de-duplication. The basics of the attack are as follows: the attacker runs in a core which shares the last level cache with the victim, and manages to share some page with it through memory de-duplication. It can either contain shared data (i.e. the tables used by the AES encryption algorithm) or shared instructions (for the attack against RSA). In the first phase of the attack (*Flush*), the attacker evicts the shared blocks from its own private cache, causing the eviction of those data from the shared cache and all the other caches. In the second phase, the victim performs some random work, bringing some of the shared data to the cache again. In the last phase (*Reload*), the attacker accesses every shared data, measuring the time it takes and it guesses which data have been used by the victim (cache hits) and which ones were not used (cache misses). From this information, the attacker extracts relevant data, such as the AES key.

4   Implementation of the AES Attack

4.1   Experimental Setup

The experimental setup was deployed on a dual-socket server featuring two Intel Xeon Gold 6138 chips with 20-cores each (hyperthreading was disabled), running at 2 Ghz. The memory hierarchy comprises 96 Gbytes DDR4 RAM, 28 Mbytes of unified L3 cache per chip (11-way associative), 1 Mbyte of unified L2 cache per chip (16-way associative) and 32 Kbytes of L1 cache per core (8-way associative). Cache line is 64 bytes. L1 TLB comprises 64 entries (4-way associative) with a page size of 4 Kbytes.

From the software perspective, we employed a Debian GNU/Linux distribution with kernel 4.9.51-1 and GCC 6.3.0. PAPI version 5.5.1.0 [14] built on top of the Linux perf_event subsystem was employed to extract performance counters information. OpenSSL version 1.1.1.b was used to implement the cryptographic algorithm, compiled with the no-asm flag when using T-tables.

4.2 AES Algorithm

The AES algorithm is iterative, so obtaining each encryption needs the execution of several rounds. In [1,7], authors develop the underlying theory of polynomials with coefficients in $GF(2^8)$ (Galois Field of order 256), which is the base for the extraction of transformation values of a single round. The round transformation lies in four steps for the first rounds (SUBBYTE, SHIFTROWS, MIXCOLUMNS and ADDROUNDKEY) and three for the last round (all but one, MIXCOLUMNS). The number of rounds will depend on the length of the key; in our case, for 128-bits key, we need 10 rounds. As stated by Daemen and Rijmen in [7], the round transformation of AES can be optimized with 4 look-up tables (T_i, $i \in 0 \cdots 3$) that contain the pre-calculated values for each of the potential inputs. This way, the encryption round is simplified to a few XOR operations and takes the form:

$$S_{i,j} = T_i[s_{i,j}^k] \oplus RoundKey_{i,j}^k \qquad (1)$$

for the main rounds, and the last round:

$$S_{i,j} = T_{(i+2)\%4}[s_{i,j}^{10}] \oplus RoundKey_{i,j}^{10} \qquad (2)$$

with $S_{i,j}^k$ the encrypted char, $s_{i,j}$ the previous state char, $k \in 1 \cdots 9$ the k-th round and s_1 is the original message (s_0) XOR with $RoundKey^0$.

4.3 Implementation of the Attack

The basis of the attack is simple: using T-Tables optimization to extract the last round key (LRK) of AES. In Sect. 2, we exposed previous algorithms for extraction of the AES key. We use the approach of [4] to break the OpenSSL 1.1.1.b AES 128 bits implementation (this library has had to be compiled with no-asm flag, so that it uses the T-Tables implementation). The attack begins by forcing the de-duplication of library pages (see Sect. 3.2). This step is mandatory so that victim and attacker can share pages of the dynamic library, hence allowing the observation of memory addresses assigned to AES tables. In order to obtain the origin of the dynamic library, we proceed by opening the library and performing a memory projection (through mmap). Proceeding this way, the KSM daemon will detect a matching in the contents of the mapped file and the loaded dynamic library, and will force the de-duplication. We have experimentally observed a delay of around 300 encryptions to unleash the de-duplication of pages. At that point, the attack can commence. The start addresses of each table are obtained by decompiling the library and determining the offset of each table with respect to its start address.

As seen in Sect. 2, there are different ways to extract the key based on the information left by the last round of encryption. In this work, we check whether a cache line[1] resides in L3 upon completion of the encryption process.

---

[1] A cache line – 64 bytes in our target architecture – can store 16 elements of a table, provided each element is stored as a 4-byte unsigned integer.

These measurements have been carried out empirically by a *Flush&Reload* technique (see Sect. 3.3) for each one of the four tables. In the following, T_j is the corresponding line of the observed table; the attack proceeds by first performing a *flush* operation of different lines of the table, followed by a random encryption request. The response to this request is then stored ($S[i]$ stores the encrypted text on the i-th encryption), together with the information that will be necessary to perform the attack: a matrix X is created and X_{ij} set to 1 if line T_j was in L3 after completing the i-th encryption, 0 otherwise.

Once these data are obtained, we proceed by searching for the most probable characters belonging to the last round key, following the pseudo-code depicted in Algorithm 1.1, that will return, for each position of the last round key, those characters with the lowest probability. Hence, we will select:

$$LastRoundKey_{i,j} = \min_{t \, \in \, 0,...,num\_encrypt} LRK_{i,j}[t] \qquad (3)$$

Once the characters of the last round key have been obtained, the last step is just an inversion of the code used by AES to obtain the last round key, and hence the initial key of the server.

Listing 1.1. Pseudo-code to obtain Last Round Key candidates.

```
1   for  t  in  0,··· ,num_encrypt
2     for  i  in  0,1,2,3
3       if  X[( i+2)%4][ t ]  == 0
4         for  j  in  0,1,2,3
5           for  l  in  0,··· ,line_elems
6             LRK_{i,j}[S^t_{i,j} ⊕ T_{(i+2)%4}[l]] + +
7           end  for
8         end  for
9       end  if
10    end  for
11  end  for
```

5   Attack Detection Using PMCs

Cache timing attacks cause an anomalously high number of L3 misses, due to the flush and reload activity; hence, measuring L3 misses is an straightforward mechanism to detect them. As explained in Sect. 2, there have been some works in this field and most of them use L3 misses.

In addition to L3 cache misses, we chose the total number of load instructions executed by the victim as a way to estimate the number of encryptions being performed by the victim, so that the ratio between both counters provides a metric that is constant for different levels of load in the victim. Thus, our detection metric is the amount of L3 cache misses (in thousands) per load instruction (1000 * L3 misses/LD instruction).

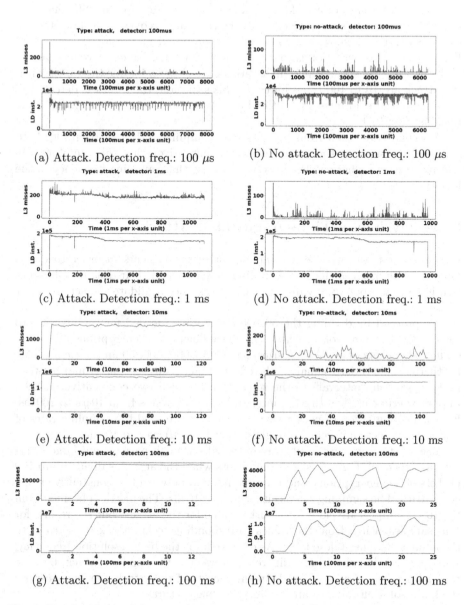

Fig. 1. Results obtained from performance counters at different sampling rates. Each one of the four rows reports the results obtained for the L3 cache misses (above) and number of load instructions (below) in the victim under attack (left) and with no attack (right). The rows correspond to the different sampling rates analyzed: $100\,\mu s$ (first row), $1\,ms$ (second row), $10\,ms$ (third row) and $100\,ms$ (last row).

Figure 1 reports the values of the chosen PMCs for the victim both in the presence and absence of attack. The experiment was repeated at different sampling frequencies, to study the effect of the sampling frequency in the detection capability. Figure 2 shows the values of the proposed metric for the results in Fig. 1. The first observation is that the selected metric is an effective mechanism to detect the attack; the values under attack are close to 1 while the values without attack are 10 to 100 times lower. In this situation, the attack is detected if, after the initial cold misses (identified as 100 ms in our experiments), the value remains close to 1.

A second conclusion from Fig. 2 is that sampling PMCs at 100 μs leads to more noisy results for the no-attack experiment. Given that this sampling rate also produces a higher overhead, we will not use that sampling frequency in the following.

6   Analysis of a Time-Fragmented Attack

In this section, we propose a complete set of experiments in order to determine if the division of the attack in discrete pieces and their distributed execution in time can potentially disguise the attack and invalidate the action of our detector.

We proceed by dividing the 50,000 encryptions needed for the attack into equally-sized groups (or "packets" in the following) of encryptions. We have evaluated packets of decreasing sizes, namely: 5,000, 500, 50 and 5 encryptions. Furthermore, in order to analyze the effect of increasing the gap (time) between packets, we call "interval" to the separation between two consecutive packets. In our experiments, we vary the interval between packets from 10 μs to 100 ms. For each combination of packet size and interval, we used the sampling rates of the previous section except the highest one: 1 ms, 10 ms and 100 ms.

The most interesting results are obtained for small packets and large intervals, as expected. Figure 3 shows the results when the attack is divided into 100 packets of 500 encryptions, and the time interval between two consecutive packets is 10 ms. The sampling rate is either 1 ms or 10 ms. The metrics obtained from the 10 ms samples are close to the usual value for the attack, but the results for 1 ms samples switch from the values corresponding to an attack (close to 1) to the no-attack values (close to 0). As expected, for the high resolution frequency, some samples do not find any difference between attack and no-attack, because they fall in the interval of time between packets of the attack. On the contrary, the low resolution samples always find the "big picture".

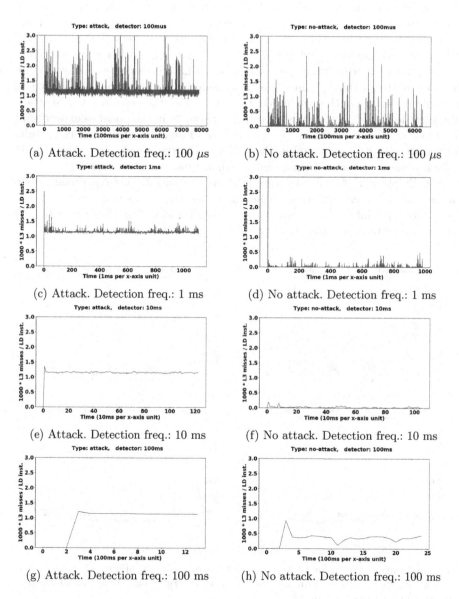

Fig. 2. Metric evaluation for attack detection at different sampling rates. Each row displays the proposed metric: 1000 L3 cache misses per load instruction in the victim under attack (left-side column) and without attack (right-side column). The four rows correspond to the four sampling rates analyzed: 100 μs (first row), 1 ms (second row), 10 ms (third row) and 100 ms (last row).

Figure 4 reports an equivalent evaluation for packets 10 times smaller, with the aim of reducing the time in which the attack can be detected. In this case, the difference between the obtained results at different sampling rates is more evident. The time interval between two consecutive packets is 10 ms and the sampling rates are 1, 10 and 100 ms. For the 1 ms sampling rate, on one hand, the no-attack experiment has higher number of L3 misses due to the separation between packets. During those intervals some cache lines are evicted due to normal functioning of the system. On the other hand, the experiment with attack also switches from low to high values of the metric as in the previous experiment. This fact can be observed in Fig. 5, which is an augmented view of the results for the attack with 1 ms sampling. It confirms that the attack can be more easily hidden from the high resolution samples than from the lower ones.

Finally, the packet size is decreased to 5 encryptions. In this experiment, when the interval between packets is longer than 1 ms the attack stops working. The results for 1 ms interval are show in Fig. 6 and they confirm that our detection metric is able to detect the attack with a sampling rate of 10 ms or bigger.

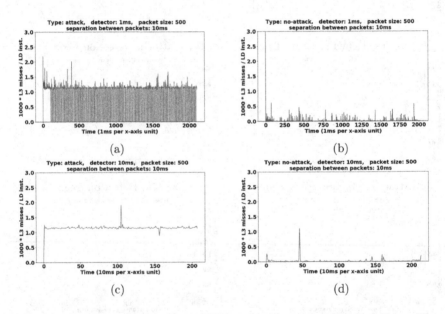

Fig. 3. Detection metric for packets of 500 encryptions with interval of 10 ms. There is an attack in the left column and no attack in the right one. The sampling rate is 1 ms (top) and 10 ms (bottom).

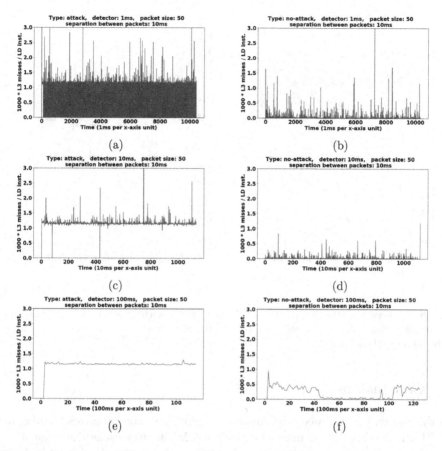

Fig. 4. Detection metric for packets of 50 encryptions with interval of 10 ms. There is an attack in the left column and no attack in the right one. The sampling rate is 1 ms (top), 10 ms (middle), and 100 ms (bottom).

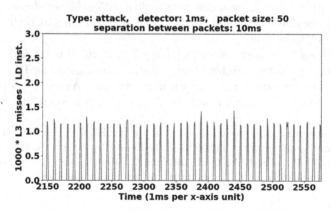

Fig. 5. Augmented view of the detection metric for the attack with packets of 50 encryptions, interval of 10 ms and sampling rate of 1 ms.

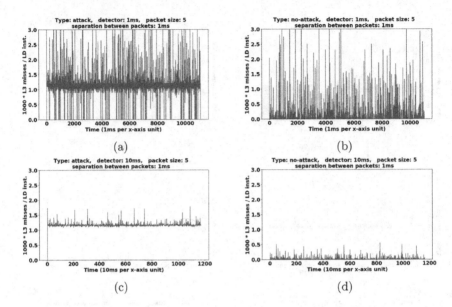

Fig. 6. Detection metric for packets of 5 encryptions with interval of 1 ms. There is an attack in the left column and no attack in the right one. The sampling rate is 1 ms (above) and 10 ms (below).

7 Conclusions

In this paper, we proposed a mechanism to protect victim processes running in multi-core servers (either native or inside a VM) against cache timing attacks by adding to the server a new detector process that monitors only the PMCs associated to the victim process. To that end, we implemented a cache timing attack against the table based AES encryption algorithm. We used 1000 L3 cache misses per load instruction as a detection metric and achieved detection of the attack for all the different sampling rates, although sampling at high frequency is worse than at lower ones.

We have tried to hide the attack dividing it into small parts and interleaving time slots with attack and without attack. Thus, sampling PMC at high frequency makes detection of the attack more difficult. Again, lower frequency monitoring (10 ms and 100 ms) results in higher detection capability.

Acknowledgements. This work is supported by the EU FEDER and the Spanish MINECO under grant number TIN2015-65277-R and by Spanish CM under grant S2018/TCS-4423. We would like to thank Samira Briongos and Pedro Malagón for their helpful comments on some details of the attack implementation.

References

1. Specification for the advanced encryption standard (AES). Federal Information Processing Standards Publication 197 (2001). http://csrc.nist.gov/publications/fips/fips197/fips-197.pdf
2. Biswas, A.K., Ghosal, D., Nagaraja, S.: A survey of timing channels and countermeasures. ACM Comput. Surv. **50**(1), 1–39 (2017). https://doi.org/10.1145/3023872
3. Briongos, S., Irazoqui, G., Malagón, P., Eisenbarth, T.: CacheShield: detecting cache attacks through self-observation. In: CODASPY, pp. 224–235 (2018). https://doi.org/10.1145/3176258.3176320
4. Briongos, S., Malagón, P., de Goyeneche, J.M., Moya, J.: Cache misses and the recovery of the full AES 256 key. Appl. Sci. **9**(5), 944 (2019). https://doi.org/10.3390/app9050944
5. Canella, C., et al.: A systematic evaluation of transient execution attacks and defenses (2018). http://arxiv.org/abs/1811.05441
6. Chiappetta, M., Savas, E., Yilmaz, C.: Real time detection of cache-based side-channel attacks using hardware performance counters. Appl. Soft Comput. J. **49**, 1162–1174 (2016). https://doi.org/10.1016/j.asoc.2016.09.014
7. Daemen, J., Rijmen, V.: The Design of Rijndael: AES - The Advanced Encryption Standard. Springer, Heidelberg (2002). https://doi.org/10.1007/978-3-662-04722-4
8. Ge, Q., Yarom, Y., Cock, D., Heiser, G.: A survey of microarchitectural timing attacks and countermeasures on contemporary hardware. J. Cryptogr. Eng. **8**(1), 1–27 (2018). https://doi.org/10.1007/s13389-016-0141-6
9. Horn, J.: Project zero - reading privileged memory with a side-channel (2018). https://googleprojectzero.blogspot.com/2018/01/reading-privileged-memory-with-side.html
10. Irazoqui, G., Inci, M.S., Eisenbarth, T., Sunar, B.: Wait a minute! A fast, cross-VM attack on AES. In: Stavrou, A., Bos, H., Portokalidis, G. (eds.) RAID 2014. LNCS, vol. 8688, pp. 299–319. Springer, Cham (2014). https://doi.org/10.1007/978-3-319-11379-1_15
11. Kumar, A., et al.: Future Intel Xeon Scalable Processors. Hotchips (2018)
12. Lyu, Y., Mishra, P.: A survey of side-channel attacks on caches and countermeasures. J. Hardw. Syst. Secur. **2**(1), 33–50 (2017). https://doi.org/10.1007/s41635-017-0025-y
13. Nguyen, K.T.: Introduction to Cache Allocation Technology in the Intel® Xeon® Processor E5 v4 Family (2016). https://software.intel.com/en-us/articles/introduction-to-cache-allocation-technology
14. Terpstra, D., Jagode, H., You, H., Dongarra, J.: Collecting performance data with PAPI-C. In: Müller, M., Resch, M., Schulz, A., Nagel, W. (eds.) Tools for High Performance Computing 2009, pp. 157–173. Springer, Heidelberg (2010). https://doi.org/10.1007/978-3-642-11261-4_11
15. Yarom, Y., Falkner, K.: FLUSH+RELOAD: a high resolution, low noise, L3 cache side-channel attack. In: Proceedings of the 23rd USENIX Conference on Security Symposium, pp. 719–732 (2014)
16. Zhang, T., Zhang, Y., Lee, R.B.: CloudRadar: a real-time side-channel attack detection system in clouds. In: Monrose, F., Dacier, M., Blanc, G., Garcia-Alfaro, J. (eds.) RAID 2016. LNCS, vol. 9854, pp. 118–140. Springer, Cham (2016). https://doi.org/10.1007/978-3-319-45719-2_6

Hybrid Elastic ARM&Cloud HPC Collaborative Platform for Generic Tasks

David Petrocelli[1,2]([✉]), Armando De Giusti[3,4], and Marcelo Naiouf[3]

[1] Computer Science School, La Plata National University, 50 and 120,
La Plata, Argentina
dmpetrocelli@gmail.com
[2] Lujan National University, 5 and 7 Routes, Luján, Argentina
[3] Instituto de Investigación en Informática LIDI (III-LIDI),
Computer Science School, La Plata National University - CIC-PBA,
50 and 120, Buenos Aires, Argentina
{degiusti,mnaiouf}@lidi.info.unlp.edu.ar
[4] CONICET - National Council of Scientific and Technical Research,
Buenos Aires, Argentina

Abstract. Compute-heavy workloads are currently run on Hybrid HPC structures using x86 CPUs and GPUs from Intel, AMD, or NVidia, which have extremely high energy and financial costs. However, thanks to the incredible progress made on CPUs and GPUs based on the ARM architecture and their ubiquity in today's mobile devices, it's possible to conceive of a low-cost solution for our world's data processing needs.

Every year ARM-based mobile devices become more powerful, efficient, and come in ever smaller packages with ever growing storage. At the same time, smartphones waste these capabilities at night while they're charging. This represents billions of idle devices whose processing power is not being utilized.

For that reason, the objective of this paper is to evaluate and develop a hybrid, distributed, scalable, and redundant platform that allows for the utilization of these idle devices through a cloud-based administration service. The system would allow for massive improvements in terms of efficiency and cost for compute-heavy workload. During the evaluation phase, we were able to establish savings in power and cost significant enough to justify exploring it as a serious alternative to traditional computing architectures.

Keywords: Smartphone · Distributed Computing · Cloud Computing ·
Mobile Computing · Collaborative Computing · Android · ARM · HPC

1 Introduction

Since their inception, x86 CPUs (Intel/AMD) and their corresponding GPUs (Intel/AMD/NVidia) were developed to solve complex problems without taking into account their energy consumption. It was only in the 21st century that, with the exponential growth of data centers, the semiconductor giants began to worry and seriously tackle their chips' TDP, optimize their architectures for higher IPC, and lower their power consumption [1–3]. If we analyze the current context, the power and

M. Naiouf et al. (Eds.): JCC&BD 2019, CCIS 1050, pp. 16–27, 2019.
https://doi.org/10.1007/978-3-030-27713-0_2

cooling bills are two of the biggest costs for data centers [4, 5]. It is for this reason that energy consumption and efficiency are two key issues when designing any computing system nowadays and it is acknowledged that accomplishing compute benchmarks must not be done so by brute force but rather by optimizing resources and architectures, keeping in mind the high financial and environmental cost of the energy required by large data centers.

Unlike the x86 processors, ARM-based chips were conceived with power efficiency as key priority from their inception, since these were oriented to the mobile devices and micro devices, which are for the most part battery-powered. Although the raw power of ARM chips is lower than that of their x86 counterparts, they are much more efficient power-wise [6–8]. At the same time, ARM-based devices vastly outnumber traditional x86 computers, so while their individual computer power might be lower, there is a very large install base with long idle periods while charging that, if properly managed, could become massive distributed data centers consuming only a fraction of the energy for the same computing power as their traditional counterparts. This distributed workload during idle time principle has been used before in x86 for collaborative initiatives such as SETI@Home and Folding@Home which use the computing power of computers that people left turned on during idle period to construct large virtualized data centers. Currently some investigators have converted this collaborative model to work on mobile devices [9–12].

After analyzing the methodology, structures, and results obtained in those case studies, our team built and evaluated a collaborative platform for HPC based on ARM-based mobile devices, following the footsteps of the previous study [8]. The platform receives data fragments and instructions from its clients via de cloud portal, generates tasks and distributes them to the worker nodes (mobile devices) for them to apply their massively parallelized computing power towards the function requested by the client. Once the task is complete, all worker nodes return their processed fragments to the central nodes and the final processed data is stored until the client requests it. In this paper in particular the implemented task was video compression using the FFMpeg library, which allows for different profiles and compression configurations to be requested by the client, as defined by the previous study [8]. The platform was evaluated through a series of performance and power usage metrics both on ARM well as x86 chips. This allowed us to make a comparative analysis between the architectures and demonstrate that it is completely feasible to offload compute heavy workloads to ARM architectures. It also allowed us to compare the power usage for the same task on ARM-based chips compared to their x86 counterparts. An analysis of a cloud-based x86 architecture (IaaS) was also performed with the objective of offering an estimation of the costs which could be recuperated if those tasks were to be migrated to the collaborative computing platform.

In order to clearly describe the different aspects of the work, the rest of the paper is organized in the following manner: Sect. 2 details the implemented protocol, technologies used, and functions developed for it. Section 3 describes the testing model and metrics used. Based on that section as well as the data collected during the experiment, Sect. 4 presents an analysis of the results. Finally, the last chapter will focus on the conclusion and future work to further explore the topic.

2 Distributed Architecture for HPC on ARM

The developed architecture implemented a hybrid model composed of (a) cloud-based resources, and (b) mobile devices. In Fig. 1 a functional diagram is presented.

The cloud resources (a) carry out the following functions:

- RESTful web server for task reception and delivery
- Admin and task management system
- Database storing statistics (time used per task and power usage)

The mobile devices (b) process the task fragments they receive in a collaborative and distributed fashion. Each node has access to the application (apk) that allows it to access the network and includes the video compression logic.

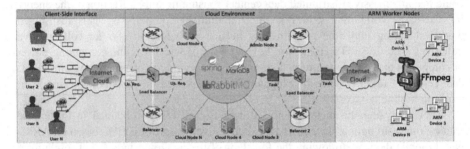

Fig. 1. Functional diagram of the collaborating computing network

2.1 Cloud Nodes

RESTful Web Servers for Task Reception and Delivery. A RESTful web app was developed through the SPRING Java framework so that both clients and nodes can communicate with the web servers in the network. A RESTController entity was developed for the purpose of establishing the connection with the distributed queuing system (RabbitMQ), with the statistics database (MariaDB), and translating the HTTP requests from the users into specific functions as determined by the roles they possess. All communication between clients, worker nodes, and servers is done through HTTP messages encoded using the JSON format, as are the task objects (Messages).

Admin and Task Management System. In order to construct an admin and task management system that was persistent, redundant, and fault-tolerant, the RabbitMQ middleware was used as a message queue-based persistence module and was integrated with the RESTful Java platform. With this tool, we implemented persistent FIFO queues and they were configured to offer high cluster availability. The messages storage uses the exchange format (JSON).

At the same time, the REST Java server implements a manual configuration model for the content of the general message queue whence worker nodes obtain their tasks. In the event of a server error, client-side issues, or execution timeout, the admin thread moved the task back from pending to available and places it back on the queue so that another worker can process it.

Database with Statistics Data Storage (Time and Power Usage). The web server also includes a link to the MariaDB database through a CRUD schema using the DAO design pattern. The database must register each executed task, which worker node executed it, the type of task, the processing time (milliseconds), and the power usage (watts). This allows us to access that information for the purpose of evaluating different architectures and derive their efficiency from the comparative analysis between their performance and power usage.

2.2 Mobile Worker Nodes

The application for the mobile nodes was developed on the native Android environment (Android Studio) and an apk was generated based on a Java codebase. The application is basically a series of modules that allow for the execution of the compute-heavy tasks to be performed without negatively impacting the OS.

The application was developed using the MVC pattern, based on activities with Java on Android Studio. A given activity defines a view that allows for the interaction with the user and visualization of the state changes made by the controller and on the flipside generates an independent worker thread which is in charge of the compute-heavy workload. This thread will continuously iterate the following process: first it connects to the REST server and downloads a task that needs to be processed, then it takes the parameters from the JSON message and applies them to the compression of the source video with the FFMpeg library which is integrated in the app as defined in the message. Once the video has been compressed, it generates a new JSON message and sends it along with the compressed video back to the server. Once the cycle is finished, it attempts to obtain another message from the queue and repeats the aforementioned process.

It should be noted that in order to evaluate and compare the different platforms, a worker node was developed in Java for the x86 architecture with the same functions as those implemented on the Android app but without a GUI since the management is done through command line.

The codebase and all the tools used in this project are available on GitHub at the following URL: https://github.com/dpetrocelli/distributedProcessingThesis.

3   Test Model and Metrics

With the goal of evaluating the performance of our prototype, we ran compute tasks composed of video compression with different sets of parameters on both types of architectures: ARM-based devices running Android and Intel x86-based devices. We aim to obtain the performance and power usage numbers for executing this task on these devices. For this purpose, the following steps were taken: (3.1) Selection of source videos; (3.2) Configuration of compression profiles and duration of video fragments; (3.3) Definition of the testing devices (ARM/x86); (3.4) Definition of the metrics to be measured; (3.5) Construction of the custom power usage meters.

3.1 Selection of Source Videos

In order to generate a test that could saturate devices of both architectures, we analyzed previous works and configurations [13, 14] and used them as a reference for this test. Based on that we selected three source videos with characteristics that were relevant for both platforms (codecs used, bitrate, compression level, frame size, aspect ratio, bits per pixel, etc.). The overview of the most important properties of each video are detailed in Table 1.

Table 1. Principal characteristics of the source videos used for compression tests

File Name	Duration	Size	Size Screen	Format	Bitrate	Comp. Prof.	Comp. lvl	FPS	Audio Codec	Kbps Audio	Sampling	Channel
3dmark-4k-120fps.mkv	2m 35 segs	487 MB	3840x2160	AVC x264	27545 Kbps	high	@L6	120	Vorbis	180	48000	2
bbb-sunflower-2160p.mp4	10 m 34 segs	605 MB	3840x2160	AVC x264	8000 Kbps	high	@L5.1	60	aac	160	48000	8
bigbuckbunny-1500.mp4	9 m 56 segs	109 MB	1080x608	AVC x264	1080 Kbps	main	@L3.1	24	aac	128	48000	2

3.2 Configuration of Compression Profiles and Video Fragment Duration

The chosen codec AVC h.264 [15] (x264 library on FFMpeg) is widely supported on current devices and video streaming platforms such as Youtube, Vimeo, Netflix, etc. The encoding profiles of x264 that were selected to run the tests mentioned in this activity are detailed below in Table 2.

Table 2. Description of the properties of compression profiles used in FFMpeg

x264 Codec	Size	Video C.	Bitrate V.	Lvl	FPS	Preset	Bframes	BF/GOP	BF Ref.	Audio C.	Bitrate A.	Sampling	Audio Ch.
High Profile													
4K	4096x2160	libx264	15600	L@5.1	60	very slow	6	3	2	ac3	512	48000	6
1080 Full HD	1920x1080	libx264	3900	L@4.1	30	slow	6	3	2	ac3	320	48000	6
Main Profile													
720 HD	1280x720	libx264	2000	L@4.1	25	medium	3	3	1	aac	320	44100	2
480	852x480	libx264	900	L@3.1	25	fast	3	3	1	aac	256	44100	2

Four profiles were selected for the two higher compression levels available on the x264 codec (main, high); these range from simple, low compression profiles, to complex, high compression ones. As such, the encoding and compression performed will be the same as the most common configurations found on video streaming platforms; which allows the public to access content in different qualities in order to match their device compatibility and connection bandwidth [16, 17].

In addition, each video is also split into 3 and 1 s fragments for each compression level. These are the recommended values for video streaming services [18, 19].

3.3 Definition of Testing Equipment (ARM/X86)

In order to evaluate the performance demands of the compression tasks, a mix of x86 and ARM-based resources were used; these were chosen based on their compute power. Both architectures have a high-end cluster (x86-i7-16gb/ARM Samsung S7), and a mid-tier cluster (x86-i5-8gb/ARM Samsung A5 2017), all of them fully support encoding/decoding of x264 videos [20]. The relevant characteristics of the equipment are listed in Table 3.

Table 3. Principal characteristics of the equipment used for compression tests

x86/ARM Cluster				
Cluster Name	Processor	Memory	Disk	GPU
x86-Clusteri7	Intel i7-4770-8x3,4Ghz	16 GB-DDR3	500GB-7200RPM	Intel HD Graphics 4600
x86-Clusteri5	Intel i5-2400-4x3,1Ghz	8 GB-DDR3	500GB-7200RPM	AMD Radeon HD 7670-480x800Mhz
ARM-ClusterS7	Exynos 8890-4x2.3 GHz+4x1.6 GHz	4 GB-LPDDR4	16 GB-SD IntStor	Mali-T880-12x650MHz
ARM-ClusterA5	Exynos 7880-8x1.9 GHz	3 GB-LPDDR4	16 GB-SD IntStor	Mali-T830-2x600Mhz

3.4 Definition of Measured Metrics

The following metrics were defined for evaluating the performance and power usage of the different architectures (x86/ARM) while performing compute-heavy tasks:

Time Metrics. Processing time per task for each architecture (IndTime).

- Minimum (MinIndTime)/Maximum (MaxIndTime)/Average (AVGIndTime)

All metrics were measured in milliseconds.

Power Usage Metrics. Register power usage metrics per task for each architecture.

- Minimum (MinWattCons)/Maximum (MaxWattCons)/Average (AVGWattCons)

All metrics were measured in watts.

It is worth mentioning that custom tools were developed for the purpose of measuring these power usage metrics. These tools are detailed in Subsect. 3.5.

Effectivity Score. The effectivity score is derived from the following metrics: the average task execution time (AVGIndTime) and power usage (AVGWattConsum) for each architecture as detailed below:

$$TimeRatio = AVGIndTime(ARM) \div AVGIndTime(x86)$$

(Determines how many times slower the ARM device is)

$$PowerUsageRatio = AVGWattCons(x86) \div AVGWattsCons(ARM)$$

(Determines how many times less power the ARM device uses)

$$EffectivityScore = PowerUsageRatio \div TimeRatio$$

If *EffectivityScore* = 1, equilibrium between architectures
If *EffectivityScore* < 1, the effectivity score is better on X86
If *EffectivityScore* > 1, the effectivity score is better on ARM.

3.5 Construction of Power Usage Devices

x86 Architecture. The x86 devices run on direct current provided by their power supply, which is fed from a standard 220 V AC mains outlet. In order to measure their power usage, a non-invasive device was built from standard micro components which consists of a current clamp wired to an Arduino controller through a headphone jack connector and a multimeter for obtaining the grid voltage.

ARM Android. While mobile devices also run on direct current, they obtain it directly from their built-in battery. Due to this, it was necessary to build a software-based power usage meter. After researching the matter [21, 22], the power usage meter protocol was integrated into the ARM worker apk through an independent thread which only runs while a task cycle is underway and does not interfere with it. This thread polls the battery sensor information from the OS every 1000 ms.

4 Experimentation and Analysis

Having defined the testing model and metrics to be measured, the experiment was executed. The analysis of the results is detailed below.

4.1 x86 Architecture Analysis

Power Usage Analysis: Taking into account the execution of the compression tasks for both architectures, the power usage of all their components (except for the monitor) required on average were: 115 W for the Intel i7 (37% higher than the CPU's TDP: 84 W) and 183 W for the Intel i5, (93% higher than the CPU's TDP: 95 W). This increase is due to the i5 having a GPU (TDP: 66 W) which aided in the assigned tasks. For this reason, the i5 cluster consumes, on average, 59% more than the i7 for all compression tasks.

Performance Analysis: Both architectures correctly completed the compression tasks (4k, HD, 720p, 480p) in the fragment configurations used (1 and 3 s) since their processing capabilities and memory (including swap) are comparatively large. The 4 k profile, both in its 1 and 3 s fragments and with the chosen compression presets, is the most complex task given that it presents a large jump in processing time compared to the other presets; this can be observed in Fig. 2. This is basically due to the bitrate of the source video, which is 4 or 5 higher than the rest, as well as the use of the highest quality profile available in the library (x264 high L@5.1) and its more in-depth precision requirement (preset: very slow). At the same time, it's clear that the i7 architecture has, on average, a 40% performance lead on its competition. Based on this it can be deduced that, the higher the task complexity, the better the i7 architecture will perform compared to the i5 architecture.

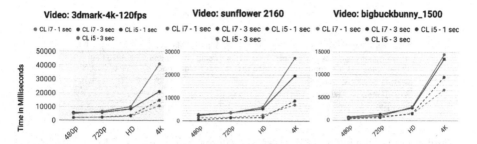

Fig. 2. x86 compression test results

x86 Platform Choice for Comparison between Architectures. The x86-i7 cluster was chosen as the one to be used for the comparison between the x86 and ARM architectures based on the performance results detailed above and the fact that the x86-i7 cluster uses on average 60% less power compared to the x86-i5 cluster.

Cost Analysis: Definition of the Cloud Architecture. For the cloud architecture, "General Purpose Av2" virtual machine instances (IaaS) on Microsoft Azure were chosen; specifically the "Standard_A8_v2" model, which contains 8 Virtual CPUs, 16 GB of RAM and 80 GB of SSD storage. It is the most similar configuration to the local i7 cluster while also being the cheapest one. Working on a Linux environment, the operating cost is 0.4$/hour (Official Azure pricing in April 2019) [23].

In order to validate the performance and costs, the same 3-video compression test suite was executed on the cloud VM but only applying the most complex profile (4K) in the smallest fragment size (1 s), with the purpose of narrowing down the results to one task in particular. The result of this testing showed that, on average, the cloud VM is 130% slower than the i7 cluster, as shown in Fig. 3.

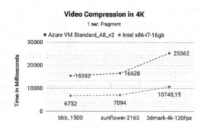

Fig. 3. x86 vs Cloud IaaS compression test results

Taking into account the requirement of compressing a 1-h video split into 1-s fragments at the most complex profile (4k), the data from the previous analysis results in cost figures of $10.15/h for the first video, $6.65/h for the second one, and $6.15/h for the third one as shown on Table 4. It should be noted that this is the estimated cost for just the processing time as it ignores any other costs arising from network usage, redundancy, traffic rules, etc. which all add further cost to cloud solutions [24].

Table 4. Azure IaaS Cost Estimates

VM: Azure Standard_A8_v2		VM Cost/hour ($)		0,4
Video Compression: 4k Video Fragment: 1 second	Comp. Time in ms per 1 Video sec.	Comp. Time in hour per 1 video hour	1 video hour Comp. Cost ($)	
3dmark-4k-120fps.mkv	25362	25,362	10,14	
bbb_sunflower_2160.mp4	16628	16,628	6,65	
bigbuckbunny_1500.mp4	15393	15,393	6,16	

Furthermore, if the first video is taken as a reference (3dmark - 487 MB - 2 min 35 s) and its file size is extrapolated to what it would be with a 1-hour duration, that would result in a file size of 11 GB. According to a study [25] from encoding.com, one of the most extensive compression and streaming cloud-based platforms on the web, they received requests to process and compress more than 6.6 PB of data between 2017 and 2018. Although the relationship is not entirely linear, handling these requests through the use of idle devices would result in significant energy as well as costs savings.

4.2   ARM Architecture Analysis

Power Usage Analysis. For video compression, the test results show that the Samsung S7 cluster consumes, on average, 3,15 W while the Samsung A5 2017 cluster consumes, on average, 2,25 W. This means that the high end platform consumes on average 40% more than the mid-range platform.

Performance Analysis. The S7 cluster is, on average, 39% faster than the A5 2017 for all the compression tasks, which matches the expected result based on their specifications. It can be concluded then that if we pair the performance and power usage metrics, they are pretty much equal in terms of performance/power.

Both platforms successfully processed the one-second video fragment tasks in all four profiles. However, when attempting the three-second video fragment tasks, the two most complex tasks failed in all four profiles (sunflower_2160 and 3dmark-4k–120 fps) and the simplest one (bigbuckbunny_1500) was successfully processed in all its four profiles. The one-second video fragment task execution times are detailed in Fig. 4 on the left side. The compression performance of the three-second video fragment tasks on the different architectures is only available for the video bigbuckbunny_1500, since the other two failed to process, and can be observed in Fig. 4 on the right side.

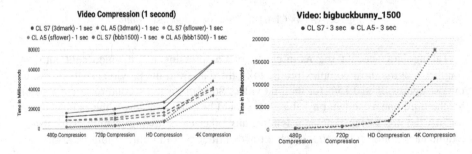

Fig. 4. ARM compression test results

The cause for failure in the three-second video fragment tasks for the most complex videos is due to the devices' inability to hold the necessary amount of data for processing that task in memory since the mobiles clusters have much less ram than their x86 counterparts (3 GB and 4 GB against 8 GB and 16 GB) and completely lack swap memory as that feature is not supported on Android.

Therefore, it is recommended that, for complex compression tasks, the video fragment size should be limited to one-second in order to fully take advantage of the mid and high-end mobile devices available in the market today.

4.3   Comparative Analysis

The performance and power usage results of the i7-x86 architecture were compared through the effectivity index against the two mobile device clusters (Samsung S7 and Samsung A5 2017), only taking into account the one-second video compression tasks that the ARM devices were able to complete successfully.

The ARM architecture, both on mid and high-end devices is between 5 and 16 times more effective compared to the x86 architecture depending upon the complexity of the task and source video. Generally, the less complex the task and source video are, the more effective ARM is. This can be observed in Fig. 5.

Furthermore, the mid-range cluster was able to process the tasks in a similar timespan to that of the high-end cluster but with a proportionally lower power usage. Thus its effectivity score is, on average, 19% higher for the bigbuckbunny-1500 video, 28% higher for sunflower-2160, and 12% higher for 3dmark-4k-120 fps.

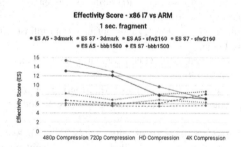

Fig. 5. Effectivity comparison of the platforms

5   Conclusions

A distributed collaborative platform was developed for the purpose of carrying out compute heavy tasks on mobile devices while they're idle and charging their batteries. It was determined through experimentation and evaluation that mid and high-end Android devices are capable of performing this type of tasks with a competitive power and cost advantage over traditional and cloud architectures. Furthermore, it was also demonstrated that the developed platform is sufficiently robust to withstand an assortment of tasks and challenges often faced by this type of applications. Lastly, it was determined that the mid-range nodes with the highest efficacy scores might not be the fastest but will always be more competitive than their more powerful counterparts owing to their lower power usage and cost.

6 Future Work

Based on the work done in this paper, the next steps to be undertaken to extend the functionalities and analysis of the platform are:

- Implement a scaling protocol for the cloud management nodes so that they can quickly respond (scale up or down) to the platform demand, as defined in [21].
- Implement other functions applicable to the massively parallel processing paradigm such as Distributed IR, machine learning, signal processing, physics simulations, pattern recognition in image and video, data clustering, etc. which allow for further testing and analysis of the capabilities of the devices and to improve the platform
- Implement a classifier of tasks and worker nodes and intelligently allocate them according to the complexity requirements and capabilities.
- Implement Credit-based incentives model for contributing idle device time to the platform.

References

1. Intelligent Machines. Intel: chips will have to sacrifice speed gains for energy saving. https://bit.ly/1PERUYu. Accessed 01 Apr 2019
2. Pandi, K.M., Somasundaram, K.: Energy efficient in virtual infrastructure and green cloud computing: a review. Indian J. Sci. Technol. **9** (2016)
3. Zaib, S.J., Hassan, R.U., Khan, O.F.: Green computing and initiatives. Int. J. Comput. Sci. Mob. Comput. **6**(7), 49–55 (2017)
4. Kania-Richmond, A., Menard, M.B., Barberree, B., Mohring, M.: A taxonomy and survey of energy-efficient data centers and cloud computing systems. J. Bodyw. Mov. Ther. **21**, 274–283 (2017)
5. Eficiencia eléctrica para Centros de Datos. https://bit.ly/2I7y3Vl. Accessed 01 Apr 2019
6. Blem, E., Menon, J., Sankar, K.: A detailed analysis of contemporary ARM and x86 architectures. In: 2013 IEEE 19th International Symposium on High Performance Computer Architecture (HPCA) (2013)
7. Pang, B.: Energy consumption analysis of ARM-based system. Aalto University School of Science Degree Programme of Mobile Computing, p. 68 (2011)
8. Petrocelli, D., De Giusti, A.E., Naiouf, M.: Procesamiento distribuido y paralelo de bajo costo basado en cloud & movil. In: XXIII Congreso Argentino de Ciencias de la Computación, XVIII Workshop de Procesamiento Distribuido y Paralelo (WPDP), pp. 216–225 (2017)
9. Arslan, M.Y., et al.: Computing while charging: building a distributed computing infrastructure using smartphones. In: 8th International Conference Emerging Networking Experiments and Technologies, pp. 193–204 (2012)
10. Gharat, V., Chaudhari, A., Gill, J., Tripathi, S.: Grid computing in smartphones. Int. J. Res. Sci. Innov. - IJRSI **3**(2), 76–84 (2016)

11. Sanches, P.M.C.: Distributed computing in a cloud of mobile phones. Faculdade de Ciências e Tecnologia, Universidade NOVA de Lisboa FCT: DI - Dissertações de Mestrado (2017)
12. Sriraman, K.R.: Grid computing on mobile devices - a point of view. In: Proceedings of the IEEE/ACM International Workshop on Grid Computing (2004)
13. Lee, B.D.: Empirical analysis of video partitioning methods for distributed HEVC encoding. Int. J. Multimed. Ubiquitous Eng. **10**, 81–90 (2015)
14. Garcia, A., Kalva, H., Furht, B.: A study of transcoding on cloud environments for video content delivery. In: MCMC 2010 Proceedings of the 2010 ACM Multimedia Workshop on Mobile Cloud Media Computing, Firenze, Italy, 29 October, pp. 13–18 (2010)
15. Linux encoding - x264 FFmpeg options guide. https://sites.google.com/site/linuxencoding/ x264-ffmpeg-mapping. Accessed 01 Apr 2019
16. Weiser, C.: Video streaming. Media Methods **38**(4), 10–14 (2002)
17. New possibilities within video surveillance (White Paper). https://bit.ly/2Vg9iJQ. Accessed 01 Apr 2019
18. Optimal adaptive streaming formats MPEG-DASH & HLS segment length. https://bitmovin. com/mpeg-dash-hls-segment-length/. Accessed 01 Apr 2019
19. Choosing the optimal segment duration. https://bit.ly/2FLeMWe. Accessed 01 Apr 2019
20. Video encoding settings for H.264 excellence. https://bit.ly/1yuCXwp. Accessed 01 Apr 2019
21. Tiwana, B., et al.: Accurate online power estimation and automatic battery behavior based power model generation for smartphones. In: IEEE/ACM/IFIP International Conference on Hardware/Software Codesign and System Synthesis, p. 105 (2010)
22. Li, D., Hao, S., Gui, J., Halfond, W.G.J.: An empirical study of the energy consumption of android applications. In: Procedings of the 30th International Conference on Software Maintenance and Evolution ICSME, pp. 121–130 (2014)
23. Configure and estimate the costs for Azure products. https://bit.ly/2UwqLk8. Accessed 01 Apr 2019
24. Mazrekaj, A., Shabani, I., Sejdiu, B.: Pricing schemes in cloud computing: an overview. Int. J. Adv. Comput. Sci. Appl. **7**, 80–86 (2016)
25. Global media formats report. https://bit.ly/2HXfSxn. Accessed 01 Apr 2019

Benchmark Based on Application Signature to Analyze and Predict Their Behavior

Felipe Tirado[1,2](✉), Alvaro Wong[2], Dolores Rexachs[2], and Emilio Luque[2]

[1] Departamento de Computación e Industrias,
Universidad Católica del Maule, Talca, Chili
ftirado@ucm.cl
[2] Computer Architecture and Operating System Department,
Universidad Autónoma de Barcelona, Barcelona, Spain
{alvaro.wong,dolores.rexachs,emilio.luque}@uab.es

Abstract. Currently, there are benchmark sets that measure the performance of HPC systems under specific computing and communication properties. These benchmarks represent the kernels of applications that measure specific hardware components. If the user's application is not represented by any benchmark, it is not possible to obtain an equivalent performance metric. In this work, we propose a benchmark based on the signature of an MPI application obtained by the PAS2P method. PAS2P creates the application signature in order to predict the execution time, which we believe will be very adjusted in relation to the execution time of the full application. The signature has two performance qualities: the bounded time to execute it (a benchmark property) and the quality of prediction. Therefore, we propose to extend the signature by giving the benchmark capacities such as the efficiency of the application over the HPC system. The performance metrics will be performed by the benchmark proposed. The experimentation validates our proposal with an average error of prediction close to 7%.

Keywords: High Performance Computing · MPI application · Performance Prediction · Performance metrics

1 Introduction

High Performance Computing (HPC) systems combine powerful hardware and software, present in clouds or clusters, used by scientists as an indispensable tool in many areas of research. The performance evaluation of these systems requires that the benchmarks subject the entire system to great stress and that they are representative of the type of workload that is executed on the machines.

In HPC, these benchmarks have followed two different approaches: The first approach consists of a set of applications and kernels, such as NAS Parallel Benchmark (NPB) [2], which aim to represent the totality of the measures of

© Springer Nature Switzerland AG 2019
M. Naiouf et al. (Eds.): JCC&BD 2019, CCIS 1050, pp. 28–40, 2019.
https://doi.org/10.1007/978-3-030-27713-0_3

performance through a set of relevant workloads. The second approach consists of a single application susceptible to the properties of the system that it considers most relevant for the typical workloads, such as the well-known High Performance Linpack (HPL) [4], which is used to classify the systems in the Top500 list [10].

When using a benchmark, it will be executed in such a way as to maximize performance, thus hiding the influence of certain properties and emphasizing the influence of other properties. For example, running HPL on very large problems makes the influence of the interconnection network negligible, causing this action to go unreported. This has the result of making it difficult to obtain an idea of application performance for different problem sizes.

Each application has data, memory structures and different arithmetic calculations, since each one tries to solve a different problem. This is reflected in the amount of memory that is needed to load the data and the type of instructions that the CPUs will compute and the memory access pattern. That is why systems exhibit different performance indices according to the applications that execute them, making it difficult to reflect, relate or select the appropriate benchmark that reflects the type of operation or the amount of data to be computed which is similar to that of the application.

PAS2P (Parallel Application Signature for Performance Prediction) [12] is a tool that allows us to analyze the dynamic behavior of the application, characterizing it in a set of phases that represent the performance of the application. With the phases, PAS2P constructs the application signature, which is defined by the set of phases which represent the application behavior at the performance level. To evaluate a system, the signature executes the phases to measure their execution times, which are multiplied by their weights, in order to obtain the total execution run time of the application.

We propose using the signature in order to create the benchmark that represents the application behavior, keeping the same memory, compute and communication requirements, as well the memory access pattern, reproducing the specific calculation and workload properties that the application has. The signature will allow us to obtain the performance behavior of each phase, where we can apply performance metrics such as the efficiency and the application execution time, obtaining results in a bounded time with high accuracy.

The performance evaluation proposal is presented in Fig. 1. Here we can observe that one of the problems is selecting the benchmark that has similar behavior to the parallel application in order to evaluate the suitable HPC system. On the other hand, our performance evaluation proposal is extracting the benchmark which represents the application performance that will be executed in order to evaluate the target machines.

In the following Section, the related works are presented. In Sect. 3, we provide general information on PAS2P methodology and Sect. 4 presents the benchmark model based on PAS2P. Section 5 provides the experimental results and Sect. 6 presents the conclusions and future work.

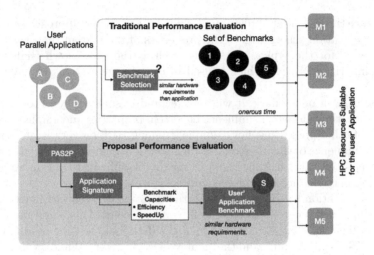

Fig. 1. Performance evaluation using benchmarks and proposal performance evaluation.

2   Related Work

In recent times, the supercomputing community has paid significant attention to three benchmarks: The HPL mentioned above, the High Performance Conjugate Gradient (HPCG) [6] and the High Performance Geometric Multigrid (HPGMG) [1]. Although HPL offers direct solutions with a computational complexity of $O(N^3)$, the two alternatives benchmarks, HPCG and HPGMG, offer iterative solutions with a linear complexity of computational calculation $O(N)$.

According to the benchmarking present in the literature, HPL and HPCG act as performance metrics and data access patterns commonly found in scientific applications, while HPGMG aims to reproduce the requirements of a specific workload class, without being clearly linked to any calculation or memory pattern, providing a balance of machine capabilities in relation to the scientific application of interest.

All the most commonly used benchmarks in HPC, in particular HPL, HPCG and HPGMG, significantly define a notion of the size of the problem, which they use as parameters to be established. But this is not enough to characterize performance, since benchmarks generally reflect the behavior of a limited set of applications, at best.

There are numerous benchmarks that represent a variety of domains. On the one hand, there is the suite of applications highlighted by the Mantevo mini-applications [5] and the Parallel NAS Benchmarks [2]. On the other hand, there is another approach consisting of a single application susceptible to the system properties that it considers most relevant for the workloads, such as HPL [4], HPCG [6]. These benchmarks are written in C/C ++ or Fortran and parallelized with MPI message passing.

Mantevo [5] presents miniapps of various kinds of scientific applications. These applications are based on the property that the performance is usually concentrated in a small subset of lines of code. This property is exploited by the miniapps, encapsulating only the most important computational operations, achieving a code smaller than the original, capturing the performance behavior of the application.

NAS Parallel Benchmarks (NPB) [2] are small application suites designed to help in the evaluation of the performance of parallel supercomputers developed by NASA. The benchmarks are based on Computational Fluid Dynamics (CFD) applications. The selection of the workload of the applications is given by five predefined classes (A, B, C, D or F). The application suite is composed of eight problems classified into five cores that mimic five numerical methods used in CFD and three simulated applications that represent a series of data calculations in complete CFD codes, which require a greater amount of resources than the cores.

HPL [4] consists of a single application composed of a single kernel. This became a point of reference in the 90s to measure the rate of execution in floating point and thus enabling the classification of supercomputers, originating the TOP500 project. HPL solved a complex system of linear equations with a complexity of $O(n^3)$. One of the main limitations of this benchmark is that it does not consider the transfer of memory or the cost in communications, which today are fundamental properties of scientific message passing applications.

On the other hand, in 2014 the benchmark HPCG [6] was developed, taking into account the limitations of HPL. It obtains a better representation of the behavior of scientific applications, making multiplications of matrix vectors in order to strongly link the benchmark with hardware memory. In addition, it uses a simple pattern and small communication messages, which make the communication time depend mainly on the latency of the interconnection network [8].

3   PAS2P Overview

Parallel scientific applications are typically composed of a set of phases that are repeated throughout the application. These phases are written in the application code using specific communicational and computational patterns. As shown in Fig. 2, PAS2P [12] identifies the application phases in a transparent and automatic way, and it generates the Application Signature, which contains the application phases (the phases which have an impact on the application's performance) and their repetition rates (weights). The Signature execution allows us to analyze and predict the application performance in an efficient way on target machines, covering approximately 95% of the total application code in 1% of the application execution time.

On the base machine, the PAS2P tool instruments each process of the application, creating a trace file. This trace, composed of hardware counters, is obtained between each MPI call. The instrumentation is performed by the MPI wrapper of the PAS2P dynamic library and the integration with the PAPI [11] library for

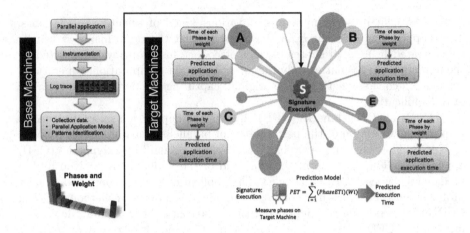

Fig. 2. PAS2P overview.

the hardware counters. Finally, PAS2P defines as an event an MPI call associated with the computational data between one MPI Call and the next one.

Once the PAS2P tool instruments the application, it analyzes the data collected in order to create a machine-independent application model. To do this, it is necessary to create a logical global clock for all processes to maintain the precedence between the events. When all the events have been ordered, The PAS2P tool creates the logical trace, where the events are inserted so as to later analyze the logical trace to extract the application phases.

To construct the signature, PAS2P instruments the application (binary); to do this, PAS2P re-runs the application using a phase table to instrument and detect where the phases occur. To predict the execution time (PET) of the application on a target machine, the equation shown in Fig. 2 is used. When the signature multiplies the execution time of each phase (PhaseETi) by its weight (Wi defined as the number of phase repetitions), the signature obtains the application execution time.

The information provided by the Performance Prediction allows us to obtain a prediction of performance measures, such as application execution time and performance metrics as computational time and the efficiency of each application phase.

4   Benchmark Based on the Application Signature

To measure the performance of an HPC system, researchers have often used a set of application kernels as benchmarks [3,4,6,7], a suite of benchmarks and mini-applications [2,5]. However, it is not always possible to characterize the performance using only benchmarks [9], due to each application having different computing and communication behavior to solve a distinct problem. This is reflected in the instructions that the CPUs will execute and the communication

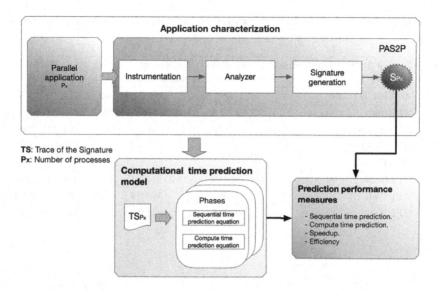

Fig. 3. Analysis and modelling the information given by the signature to obtain performance metrics.

and memory access pattern present in the application. That is why we proposed to use the signature as a benchmark. The signature uses the application code to predict whether the performance will be the same as the application. In other words, this means same compute, same communication messages and same memory access pattern. As we have said before, a very important characteristic of the benchmarks is the bounded time in which they execute, a characteristic which we have transferred to the signature. With this advantage, we guarantee that the quality of the performance results are in a bounded time.

In order to provide this capacity to the application signature, we need to analyze how we can apply the performance metrics in each phase. Each phase represents parallel code between two MPI communications and each phase scales independently from the others. This process is called application characterization (AC), as is shown in Fig. 3.

The execution of the signature allows us to extract information about the behavior of each relevant phase from the application. This information is stored in one trace file per process, as shown in Table 1, called the Signature Physical Trace (TSP_x). The TSP_x groups the information of the application in phases with its respective degree of repeatability (weight), as well as providing information from the source or destination of each message, the type of MPI primitive, the computational time between each MPI communication, the number of instructions, cycles and cache MISSES (L1 or L2).

By using TSP_x, it will be possible to model the behavior of the phases, which provide information on the application processes such as computational time, number of instructions, number of cycles and cache misses, along with the

Table 1. Information of the phases (process 1) for application N-Body with 360,000 particles.

Source	Type of MPI primitive	Destination	Computational time	Number of instructions	Cycles	Cache misses (L2)
PHASE 0 WEIGHT 289						
1	MPI_Irecv	0	3725272115	3923734167	5911572490	1107574
1	MPI_WaitAll	0	70532	62343	85415	7582
1	MPI_ISend	2	17165	412	1271	30
PHASE 1 WEIGHT 10						
1	MPI_Irecv	0	3643785450	3923734084	5778196900	1117380
1	MPI_WaitAll	0	176601	1766014	222832	3021
1	MPI_ISend	2	17687	412	1533	25

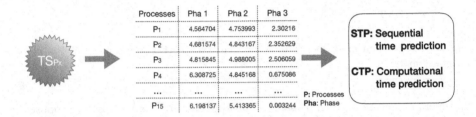

Fig. 4. Signature information: computational time of each phase per process.

weight of each phase W_m. With this information, we obtain the computational time prediction (CTP) of the executed phases, which we use to calculate the sequential time prediction (STP), as shown in Fig. 4. This stage is called the Computational Time Prediction Model (CTPM).

To predict sequential time (STP) on a parallel computer, we use the Eq. 1, for which we perform the sum of the computation time of each process of the phase $(pct_{i,p})$ and multiply it by its weight, W_i in all phases i, obtaining an

Table 2. Information obtained by executing the application signature.

CG Class D, 128 processes					
Phase ID	Compute time (sec)	Average compute time (sec)	Weight	Total compute time prediction (sec)	Representative compute time prediction (sec)
0	0.47	0.004	5024	2361.28	20.09
1	11.15	0.087	5023	56006.45	437.00
2	10.68	0.083	200	2136.00	16.60
3	1.63	0.013	199	324,37	2.58

approximation to the sequential execution time of the application. On the other hand, if we add the average computational time of each process, $\overline{x_i}$, and multiply it by its weight, W_i, in each of the phases, we obtain an approximation of the computational time of the executed application, Eq. 2, where m is the number of phases.

$$STP = \sum_{i=0}^{m}(\sum_{p=0}^{P_x} pct_{i,p} * W_i) \tag{1}$$

$$CTP = \sum_{i=0}^{m} \overline{x_i} * W_i \tag{2}$$

We exemplify the calculation of STP and CTP, which we show in Table 3. It shows two phases created by the signature when executing the LU application class D with 600 iterations in 128 processes. Each phase contains the compute time of each process expressed in seconds, as well as its weight W_i, the sum of computational time $\sum_{p=0}^{P_{127}} phase_i$, the compute average $\overline{x_{phase_i}}$ and the standard deviation of computational time σ. The STP value is obtained by the sum of the multiplication of the total computational time by the weight w_i for each phase. The CTP value is obtained by multiplying the compute average by the weight in each phase and then the sum of each one of the values.

In Table 2 we show the information obtained by the execution of the CG signature with 128 processes and workload class D with 200 iterations. In the same Table, we show the ID phase, the total computational time (all processes), the average computational time value and the weight of each phase. In the same

Table 3. Exemplification of LU application computational times with 128 processes.

Processes	Phase 0	Phase 1	
P0	0.69393	1.09571	
P1	0.71858	1.14146	
P2	0.73267	1.14258	
P3	0.72727	1.13940	
...	...	...	
P127	0.72588	0.91355	
Weight (W)	599	598	
$\sum_{p=0}^{P_{127}} phase_i$	105.904	130.353	$\sum_{i=0}^{1}(\sum_{p=0}^{P_{127}} phase_i * W_i) = 141388.350seg. = STP$
$\overline{x_{phase_i}}$	0.837	1.018	$\sum_{i=0}^{1} \overline{x_{phase_i}} * W_i = 1104.597seg = CTP$
σ	0,048	0,053	

way, we show the prediction of the total computational time, as well as the prediction of the representative computational time in each phase.

The last stage, CTPM, will allow us to numerically obtain the performance measures of the application that the user executes. STP, as previously mentioned, is obtained if we sum all the phases from the prediction of the total computing time. The value of STP for the CG application represented by the phases shown in Table 2 is 60828.1 s. Likewise, when summing the prediction of the computational time representative of all phases, we obtain CTP. The CTP value for the CG application shown in Table 2 is 476.27 s. These two times will allow us to obtain performance measures such as speedup and efficiency. These measures are specific to the executed parallel application and reflect its computing and communications behavior, allowing it to have a performance index associated with the executed application.

5   Experimental Results

Throughout this section, we will show the results of the measures obtained using the benchmark based on the application signature. In order to validate the prediction results, we compare the obtained performance time with the real execution time of the application.

In order to validate the experimental results, a set of scientific messages passing applications has been selected. Suites with different communication and compute patterns such as NAS parallel benchmarks, Mantevo and the Nbody application have been tested along with the application to be able to analyze their efficiency and execution time in comparison and therefore select the one that run better and validates the results. The four experimental applications are described in the Table 4.

Table 4. Application description.

Application	Description
MiniMD [5]	It is an application of the Mantevo suite on molecular dynamics (MD), it uses the spatial decomposition MD, where the processors of a cluster have subsets of the simulation problem
LU (Lower-Upper Gauss-Seidel solver) [2]	It is an application of the NAS suite, dealing with fluid dynamics. It solves flows in a cubic domain
CG (Conjugate Gradient) [2]	It is an application of the NAS suite that uses the inverse power method to find an estimate of the largest eigenvalue of a symmetric sparse matrix using the conjugate gradient method as a subroutine to solve the systems of linear equations
N-Body	It is an application that simulates the interaction of a dynamic system of particles under the influence of different physical forces, such as gravity

Table 5. Cluster characteristics.

Cluster	Characteristics
DELL	AMD Opteron^{TM}6200 1.60 GHz, 8 nodes (512 cores), 256 GB RAM per node (2048 GB total memory), Interconnection Infiniband QDR

Table 6. Benchmark time results compared to the whole application.

Procs.	Application		Benchmark		Error	
	Execution time (sec)	Computational time (sec)	Predicted execute time, (PET) (sec)	Predicted computational time, (PCT) (sec)	PET (%)	PCT (%)
Application: miniMD						
16	1687.11	1607.21	1509.69	1426.66	−11.75	−12.66
32	839.53	810.15	751.11	726.43	−11.77	−11.52
64	532.14	489.01	463.05	423.59	−14.92	−15.44
128	279.74	245.69	251.44	245.69	−11.25	−12.16
Application: CG						
16	6825.80	6601.83	6840.98	6601.83	0.22	−0.50
32	2360.95	2257.26	2359.94	2243.03	−0.04	−0.63
64	1407.04	1222.85	1379.23	1216.74	−2.01	−0.50
128	757.07	483.41	738.57	475.438	−2.50	−1.67
Application: LU						
16	11735.83	11101.71	12037.17	11075.91	2.50	−0.23
32	5464.36	5195.83	5447.42	5175.61	−0.31	−0.39
64	2647.85	2382.46	2750.72	2361.66	3.74	−0.88
128	1315.88	1110.75	1380.00	1110.75	4.65	−0.55
Application: N-Body						
16	1849.01	1844.83	1675.35	1700.25	−10.37	−8.50
32	927.34	821.15	864.69	860.48	−7.25	−7.05
64	469.50	463.41	446.79	446.87	−5.08	−3.70
128	397.22	386.76	410.92	364.42	3.33	−6.13

For the execution environment, a DELL machine whose characteristics are described in Table 5 was used. For the experimentation set, four different executions were performed for each application in accordance to the number of processes to be executed: 16, 32, 64 and 128 processes. N-Body was executed with a workload of 360000 particles, partitioned into the number of defined processes. For the MiniMD application of the Mantevo suite, we arranged a workload of $192 \times 192 \times 192$ with 500 iterations. The CG and LU application of the NAS suite was executed with a Class D workload with 200 and 600 iterations respectively.

The Table 6 shows the results we obtained from the executions. We used 1:1 mapping (one process per core) having a maximum of 128 cores per application. Therefore, if each node of the Dell cluster has 64 cores, when we run the application with 128 cores we are actually using two nodes. The average percentage error in the execution time of the benchmarks was 4.70%. The maximum value was 14.92%, which is below the whole application value, obtained by the miniMD application executed with 64 processes. On the other hand, the average percentage of error in computing time was 5.15%, and a maximum value of 15.44% according to the value of the whole application, obtained by the miniMD application with 64 processes. The MiniMD application obtained these results because of its distinct behaviors for different groups of processes, making the PAS2P method have a low representativeness of the application.

The Table 6 shows the scalability of the applications. It can be appreciated that the CG and LU applications have a superlinear behavior when they are executed with 16 and 32 processes. This is due to the mapping we used, which favors the use of the second level cache memory. In Fig. 7, we can observe the sublinear behavior of the applications when changing the mapping of the processes.

Our proposal allows us to obtain a prediction of the sequential time of parallel message passing applications, as seen in Fig. 4, thus, allowing us to obtain performance measures that are commonly used such as speedup and efficiency. Figure 5 and Fig. 6 show the prediction of both metrics with the mapping previously used to avoid the superlinearity of the results, with an average percentage error of 6.1% for speedup and 7.6% for efficiency. As seen in both figures, the results show a similar behavior to that of the real application.

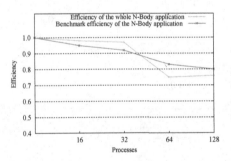

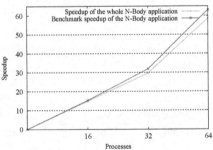

Fig. 5. Benchmark efficiency and whole application efficiency.

Fig. 6. Benchmark speedup and whole application speedup.

Running the signature instead of the full application has two important benefits. For instance, the prediction of the execution time and the computing time. It also provides extensive information concerning each phase of the application to model its behavior allowing, for example, the prediction of the sequential time of the application. The sequential execution time is often impossible to achieve, due to the amount of memory that a sequential process needs to be executed and the time involved in its execution. Figure 8 shows the bounded time in which the signature was executed versus the execution time of the entire CG application. As seen, the signature executed up to 80 times faster compared to the application, thus obtaining measurements that will allow us to evaluate the performance of the application with a low error rate.

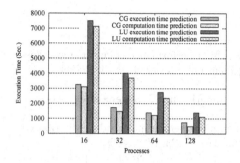

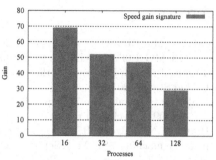

Fig. 7. Prediction of execution time with processes mapping.

Fig. 8. Speed gain of benchmark on the CG application.

6   Conclusion and Future Work

We validated the proposed benchmark based on the application signature as follows. We gave the signature the same functionality a benchmark has in order to evaluate a system. To achieve this goal, we begin by analyzing the computational behavior on each phase to predict the sequential time that the application would have. In this way, we calculate and predict computational time, speedup and efficiency using the proposed benchmark. The performance measures obtained allowed us to detect inefficiencies without executing the application completely, since the information was obtained directly from the benchmark, thus achieving a bounded execution time with a low margin of error.

Benchmarks that are based on application signatures obtain performance measures of a specific MPI application on a specific machine, without having to depend on a suite of applications or a specific application that characterizes the overall performance. In other words, what we are proposing is a benchmark adapted to the MPI application that the user wants to execute.

Currently, the signature is built using checkpoint libraries. If the user or system administrator wants to evaluate performance in a different system, it is necessary to transport the signature to the new location. One of the disadvantages of using the checkpoint mechanism is the size it achieves. The more processes the application has, the larger the size of the checkpoints we have to save. For future work, we are analyzing how to detach the checkpoint from the signature. Being portable is another important characteristic of benchmarks, we are working on the creation of compute models that include the characterization of memory access pattern as it is an important factor in the performance impact on the execution time.

Acknowledgments. This research has been supported by the Agencia Estatal de Investigación (AEI), Spain and the Fondo Europeo de Desarrollo Regional (FEDER) UE, under contract TIN2017-84875-P and partially funded by a research collaboration agreement with the Fundacion Escuelas Universitarias Gimbernat (EUG).

References

1. Adams, M., Brown, J., Shalf, J., Van Straalen, B., Strohmaier, E., Williams, S.: HPGMG 1.0: a benchmark for ranking high performance computing systems. Technical report, Lawrence Berkeley National Laboratory (LBNL), Berkeley, CA, United States (2014)
2. Bailey, D.H., et al.: The NAS parallel benchmarks. Int. J. Supercomput. Appl. **5**, 63–73 (1991). Technical report
3. Brown, P.N., Falgout, R.D., Jones, J.E.: Semicoarsening multigrid on distributed memory machines. SIAM J. Sci. Comput. **21**(5), 1823–1834 (2000)
4. Dongarra, J.J., Luszczek, P., Petitet, A.: The LINPACK benchmark: past, present and future. Concur. Comput. Pract. Exp. **15**(9), 803–820 (2003)
5. Heroux, M.A., et al.: Improving performance via mini-applications. Sandia National Laboratories, Technical report SAND2009-5574, 3 (2009)
6. Heroux, M.A., Dongarra, J.: Toward a new metric for ranking high performance computing systems. Sandia National Laboratories Report, SAND2013-4744 (2013)
7. Hoisie, A., Lubeck, O., Wasserman, H.: Performance and scalability analysis of teraflop-scale parallel architectures using multidimensional wavefront applications. Int. J. High Perform. Comput. Appl. **14**(4), 330–346 (2000)
8. Marjanović, V., Gracia, J., Glass, C.W.: Performance modeling of the HPCG benchmark. In: Jarvis, S.A., Wright, S.A., Hammond, S.D. (eds.) PMBS 2014. LNCS, vol. 8966, pp. 172–192. Springer, Cham (2015). https://doi.org/10.1007/978-3-319-17248-4_9
9. McCalpin, J., Oakland, C.A.: An industry perspective on performance characterization: applications vs benchmarks. In: Proceedings of the Third Annual IEEE Workshop Workload Characterization, Keynote Address, September 2000
10. Meuer, H., Strohmaier, E., Dongarra, J., Simon, H., Meuer, M.: Top 500 list (2012)
11. Terpstra, D., Jagode, H., You, H., Dongarra, J.: Collecting performance data with PAPI-C. In: Müller, M., Resch, M., Schulz, A., Nagel, W. (eds.) Tools for High Performance Computing 2009, pp. 157–173. Springer, Heidelberg (2010). https://doi.org/10.1007/978-3-642-11261-4_11
12. Wong, A., Rexachs, D., Luque, E.: Parallel application signature for performance analysis and prediction. IEEE Trans. Parallel Distrib. Syst. **26**(7), 2009–2019 (2015)

Evaluating Performance of Web Applications in (Cloud) Virtualized Environments

Fernando G. Tinetti[1,2(✉)] and Christian Rodríguez[1]

[1] III-LIDI, Fac. de Informática, UNLP, La Plata, Argentina
{fernando, car}@info.unlp.edu.ar
[2] CIC Provincia de Bs. As., La Plata, Argentina

Abstract. Web applications usually involve a number of different software libraries and tools (usually referred to as *frameworks*) each carrying out specific task/s and generating the corresponding overhead. In this paper, we show how to evaluate and even find out several configuration performance characteristics by using virtualized environments which are now used in data centers and cloud environments. We use specific and simple web software architectures as *proof of concept*, and explain several experiments that show performance issues not always expected from a conceptual point of view. We also explain that adding software libraries and tools also generate performance analysis complexities. We also shown that as an application is shown to scale, the problems to identify performance details and bottlenecks also scale, and the performance analysis also requires deeper levels of details.

Keywords: Performance monitorization · Web applications performance · IaC (Infrastructure as Code)

1 Introduction

As web applications and services have grown in functionality and scale, operation teams have been adopting different practices to achieve automation. Their main goal is to be up to date with development teams that have evolved much more quickly than operations. Besides, the operation teams necessarily focus on scalability problems that arises when their sites acquire popularity and the corresponding large requirements generate system failures and/or unacceptable response times. Failure in scaling up the computing resources (i.e. under-provisioning of resources) implies losing quality of service and, possibly, making a website, application or service, unavailable. Virtualization and cloud environments have provided successful tools and solutions for scaling up computing resources, but oversizing resources (i.e. over-provisioning of resources) also implies oversizing costs.

Elastic cloud computing environments claim to be appropriate for dynamically provisioning and de-provisioning resources. Furthermore, elastic cloud computing environments follow the "utility computing" model [12], and the its corresponding "pay-as-you-go" billing model, which turns to have the best cost/benefit relationship. However, depending on specific scenarios, it is hard to know how much and when scaling up or down, because it is almost completely web (site or application or service) dependent.

M. Naiouf et al. (Eds.): JCC&BD 2019, CCIS 1050, pp. 41–50, 2019.
https://doi.org/10.1007/978-3-030-27713-0_4

DevOps [3, 9] and SRE (Site Reliability Engineering) [10] practices emerged and established along the last ten years. As a consequence, IaC (Infrastructure as Code) [7] frameworks became the recommended way to simplify automation and bring resilience to infrastructures, being on-premises or cloud based data centers [4, 13]. Moreover, IaC adoption simplifies migrations from/to on-premises and cloud based datacenters and even build hybrid solutions.

Specifically related to scalability, applications must be implemented with some guidelines in mind. The Twelve Factor App Methodology [6] is a suitable starting point to adhere. However, application scalability depends on many domain-specific details, and there is not a single recipe to achieve acceptable/good results. We are going to divide the problem by services, each with different problems and options to scale

- Web application:
  - Stateless designs are scalable. State can be moved outside the application, using storage services like filesystem, databases or NoSql store engines as Memcached or Redis, among others.
  - Stateful applications can be scaled using sticky sessions. This approach is not recommended, but is preferred than no scalability.
  - The Twelve Factor App Methodology [6] enumerates best development practices to achieve scalability.
- Shared file system: not every shared file system can be scaled. Integrity is a must in some scenarios, but not for others. A shared file system can be a solution to grow, but availability becomes an issue depending on specific implementations (e.g. NFS: Network File System).
- Database: ACID (Atomicity, Consistency, Isolation, and Durability) database transaction properties is one of the main problems in a cluster of database engines. It depends on the DBMS engine to support a clustered environment or not. Some solutions involve multiple slaves and a master server. Only the master server carries out update queries, and slaves and master can be balanced to carry out read only queries. Some specific database load balancers can be used.

Web applications *tend to be scalable*, but some problems emerge when state is maintained outside the application. In this case, the whole software architecture relies in a third-party service that is not easy to implement in a scalable way as is the case of databases or shared file systems. In this paper, we are going to analyze web server configurations scalability considering a dynamic application server behind a reverse proxy to understand where the bottlenecks are, and which configuration do its best considering scalability. More specifically, we will try to provide some insight for the analysis of popular web applications, focusing in how a reverse proxy works and identifying what kind of content is served, identifying the requirements slowly served which eventually make the whole application unavailable to end users.

The rest of the paper is organized as follows. We define some important terminology and the underlying problems to which some terms are referring to. In Sect. 3 we show a simple experiment defined to show that some performance problems are found in details which are sometimes hidden or non-properly identified. Section 4 focuses horizontal web application scalability and performance evaluation. Finally, in Sect. 5 we outline some conclusions from the work presented in this work as well as our guidelines for the future work in this area.

2 Defining Terms and Problems

There are plenty of development languages, software libraries, and frameworks combined/configured in software architectures for building web applications. And web application architectures have been evolved and redefined, from a monolithic or single tiered web application, to the popular three tiers architecture, service based, or even microservices patterns architectures. Each architecture can be implemented by the number of available languages. Moreover, there are frameworks to easily develop applications following standards and so called *best practices*.

We will define some terms to better understand the context as well as specific details of our work. Web applications provide content that can be served:

- **Statically:** this content usually does not suffer any delays when served, and most of the times can be cacheable.
- **Dynamically:** depending on the requirement, the corresponding reply include delays of computing requirements and third party services (e.g. web services or database queries) used to build a response. Dynamically defined replies do not always can be cached.

Serving dynamic content requires more processing (and its corresponding delay time from the clients' point of view) and required resources than serving static content. At this point it is necessary to define and differentiate from one another application servers and (static content) web servers:

- An application server generates dynamic content as well as services related to a web application. Usually, (web) application servers are more complex than static web servers, and the "extra" complexity usually makes application servers slower than static web servers. As more time is required to reply requests, there are stronger limits to concurrency.
- Web servers commonly provide static content as assets, files, images, etc., and in some cases they are used to implement reverse proxies and even content delivery networks [11].

In this context, overcoming a limit for concurrency is directly related to scaling [1], i.e. the way in which more resources are available for processing, and in this specific case: for the application servers [2, 8]. Vertical scaling is related to hardware, i.e. resources are provided almost directly by the available hardware. Horizontal scaling, on the other hand, is usually related to services, provided by servers on hardware. Thus, horizontal scaling is usually cheaper in terms of required hardware and amount of work/configuration. Also, horizontal scaling is specially fitted to cloud environments, where the hardware is virtualized and (new and/or more) services are relatively easy to be deployed.

Application servers are naturally related to programming languages because programming is required by each specific application. Depending on each development/programming language, there are different application servers:

- Java application servers: Glassfish, JBoss EE, Apache Tomcat, etc.
- PHP: Apache with PHP module imposes the Multi Processing Module, a non-threaded, pre-forking web server. Alternatively, PHP-FPM can be used as application server, and use a web server (e.g. Apache or nginx) that reverse proxies' HTTP requests using the FastCGI protocol.
- Ruby: Unicorn, Puma, and Passenger are the most popular ruby application servers. Each one depends on a web server, usually nginx is the best companion to each case.
- Python: Gunicorn and Daphne are popular python application servers. They implement WSGI (Web Server Gateway Interface) and ASGI (Asynchronous Server Gateway Interface) protocols to communicate with reverse proxies in front of them. Nginx is generally the chosen reverse proxy.

The above list enumerates several of the most popular web development languages and their corresponding application servers. In this work, we will use Apache with PHP module to emulate an application. We will handle the PHP application for experimentation purposes, e.g. controlling/defining its response time. Our approach will be to include a delay time, emulating a third-party time service, and statically (e.g. as a parameter) set in a specific amount of time.

We will make several experiments by building different service and software architectures. In each experiment, we will simulate traffic/a pattern of requirements for the analysis of results by recording reply time and/or errors (e.g. timeouts). We will take advantage of IaC tools, i.e. the same tools currently used for maintaining on production websites and applications.

From an operations point of view, we have many alternatives for implementing our experiments: virtualization based on Virtualbox, VMWare, Hyper-V or Xen, or even cloud provided PaaS (Platform as a Service). Although all of them are suitable implementation tools, their use implies to develop shell scripts, playbooks, and/or receipts in order to take advantage of idempotent framework custom scripts such as Ansible or Chef. Instead, we are going to implement our experiments using docker and docker-compose tools which will allow (easy) versioning, replication, and scalability.

3   Simple Experiments: Where Are the Problems?

We will use a PHP script specifically designed to run in a fixed amount of time by sleeping the script by two seconds before generating the reply to the corresponding request. Besides, we configure the apache server for handling only two concurrent requests. The combination of the apache web server configuration and the PHP script request handling imposes some restrictions on how this web architecture works. With this fixed time and resources restrictions, we expect the following behavior: (a) Only two requests can be served concurrently, (b) Each request will have a delay of 2 s, and (c) We expect to serve 60 requests per minute without any errors.

We test this architecture using the Apache Benchmark tool, for different concurrent requests configurations and a total of 600 requests. The experiment for 600 requests with a concurrency level of 3 requests is made by executing

```
ab -1 -c3 -n 600 http://localhost:8080/
```
And the following summary is obtained:

```
Concurrency Level:       3
Time taken for tests:       600.459 seconds
Complete requests:       600
Failed requests:         0
Total transferred:          148200 bytes
HTML transferred:        31200 bytes
Requests per second:     1.00 [#/sec] (mean)
Time per request:           3002.296 [ms] (mean)
Time per request:           1000.765 [ms] (mean, across all concurrent
requests)
Transfer rate:              0.24 [Kbytes/sec] received
Connection Times (ms)
                  min     mean[+/-sd]   median    max
Connect:           0        0    0.2        0         3
Processing:  2001  2999 1000.0     2007      4005
Waiting:           0    998 1000.1            6     2004
Total:        2001  2999   999.9     2007      4005
```

As another example, the experiment for 600 requests with a concurrency level of 10 requests is made by executing

```
ab -1 -c10 -n 600 http://localhost:8080/
```

And the following summary is obtained:

```
Concurrency Level:       10
Complete requests:       600
Failed requests:         0
Total transferred:          148200 bytes
HTML transferred:        31200 bytes
Requests per second:     1.00 [#/sec] (mean)
Time per request:           10007.641 [ms] (mean)
Time per request:           1000.764 [ms] (mean, across all concurrent
requests)
Transfer rate:              0.24 [Kbytes/sec] received
Connection Times (ms)

                 min   mean[+/-sd] median    max
Connect:           0       0     0.2        0         2
Processing:  2002  9941   629.9  10007     10012
Waiting:           1    7940   629.9     8007      8012
Total:        2002  9941   629.8  10007     10012
```

Figure 1 shows the results for different number of concurrent requests, and as concurrency grows, the clients experience a slower response because of the limitation of two simultaneous clients configured at the web server.

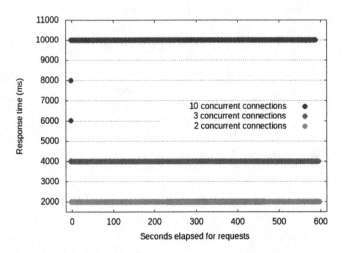

Fig. 1. Simple Server configuration, 2 concurrent requests limit set at the Apache web server.

From the point of view of the PHP server, more concurrent requests should imply less average reply time (within the limits of PHP server computer resources such as RAM size). For 2 concurrent requests, server throughput is only one request per second, as shown in Fig. 1, because each request will be replied in 2 s, and they are concurrent. In the example, when more than 2 concurrent requests are received, only the first two are handled as expected, all the other requests are queued at the Apache web server (not the PHP server). As more concurrent requests are made, the average reply time will proportionally grow, because the (low) fixed number of concurrent requests handled by the Apache server. Clearly, the problem is not the PHP server (maybe the *traditionally* first "candidate" for optimization and/or performance analysis), but the Apache web server configuration.

4 Looking at More Complex/*Real* Problems

As explained in the previous section, limits in the number of concurrent connections configured in the web server may result in increasing response time once those limits are exceeded. In general, each request implies to acquire and use an amount of resources by the application server, like memory, CPU, or even IO. This resource consumption is the main factor to consider when calculating how many requests to serve concurrently. When resources usage get near the physical limits, we must approach upward scalability. At this point is when horizontal scaling is usually adopted as a general solution. Experimentally, it is possible to horizontally scale up the above architecture and test it with the same tools to compare results. The scale down problem is rather analogous from the experimental point of view.

We have set an application server which can be horizontally scaled by means of a standard load balancer, as schematically shown in Fig. 2. It is worth noting that using a load balancer for horizontal scaling is usually easy: install the corresponding tool/service and configuring a few parameters, such as an *upstream timeout*, the maximum waiting time for a backend server reply.

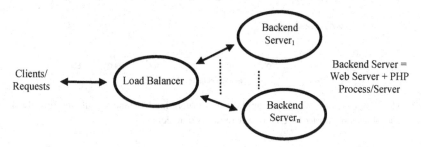

Fig. 2. Horizontal scale up of web application server.

Figure 3 shows the results obtained with a load balancer configuration timeout of 15 s from each upstream (Apache + PHP server, or *backend server*) to obtain a response. We are testing the worst scenario from the previous experiments, i.e. that with 10 concurrent requests, and scale up the backend servers from one backend server to two and ten backend servers.

As Fig. 3 shows, using only one backend server there is no performance difference from that shown of Fig. 1. For two instances of the backend, the requests are served at about half the time in average: 50% being served in 4 s, and the other 50% served in 6 s. With ten backend server instances, the 600 requests are served in approximately 2 min, being all requests replied in 2 s in average, the minimum time each request would be served.

Horizontal scale up is easily implemented with a load balancer, as shown in the previous experiment. However, adding a load balancer also adds new details, including configuration parameters and monitorization data. More specifically, the load balancer is defined with a timeout of 15 s for each upstream backend server, as detailed above. As this timeout is lowered, more non-200 status codes HTTP responses will be generated directly from the load balancer to the clients. In the experiments shown in Fig. 2 above all the requests had a 200 HTTP response code, i.e. every request was successfully replied. Reducing the timeout below 10 s, the load balancer replies a fraction of the requests with non-200 HTTP response codes, more specifically: 504 and 499 response codes. The specific fraction replied with error depend of the specific timeout set at the load balancer, as expected. The load balancer timeout value must be fine-tuned, knowing how much time each upstream request will last and the number of concurrent requests each upstream would successfully handle to achieve a proper web application response behavior.

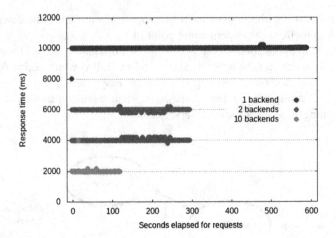

Fig. 3. Horizontal scale up of web application server: load balancer with multiple backend servers, each handling up to 10 concurrent requests/connections.

The full software configuration and running experiments of the previous section as well as the ones in this section can be found in a repository at [14]. Even when a lot of "what-if" questions can be analyzed by statistic methodologies and queueing theory in particular [5], having an experimentation environment provides several advantages. In the extreme case the real application can be used along with real data collected from on production site/s. Besides, some simple changes in the environment would provide direct results, just as those described before: reducing the load balancer upstream timeout results in a proportionally large number of non-200 HTTP response codes for the corresponding HTTP requests.

5 Conclusions and Further Work

We have shown in simple scenarios and experiments several details of web application server analysis of performance and scalability. We also have set our experiments with easy replication and control version by means of currently IaC (Infrastructure as Code) tools. More specifically, we have shown that performance reply and also timeout errors may be related to standard tools such as web servers and load balancers instead of the application specific code/service (the PHP code in our example). We have also shown the effects of horizontal scale up the application server and the corresponding perfor-mance enhancement. Even when we have focused the problem at the load balancer, in a real application there are three sources of delay time and possible problems for replying each client request:

- The load balancer process, which depends on its configuration, e.g. the timeout defined for the upstream backend server/s in the example we have given above, and the number of upstream backend server processes.

- The web server, part of the backend server process, which also depends on its configuration, e.g. the number of allowed concurrent requests in the experiments/examples given above.
- The application itself, which in this case is a simple PHP process with a predefined delay, but maybe as complex as the problem/business requires, including specific source code and third-party services such as databases.

All the experiments and details explained above are extremely important in a real web application, because it is almost impossible to determine response times with a new release of an application. At this point, statistics tools, monitoring, and observability must be driving the limits as part of the testing phase of a new web application. And, once in production, almost the same monitoring process should be carried out for at least aiding the runtime performance evaluation process. Even when we have shown scale up examples, similar tests/experiments can be defined and carried out for scale down resources in order to avoid over-provisioning of resources and the corresponding extra costs in cloud environments.

The tools and web application architectures are widely available and used in current in-production web applications. Furthermore, deploying new web application versions without the proper performance experimentation usually ends up in failure or over-provisioning of resources and the corresponding extra cost in either cloud services or hardware in a data center.

One of the most interesting work to be carried out as a next step is focused in defining a methodology for monitorization and triggering process for automating at least some scaling (up and down) processes. Current alarm triggers are usually defined per monitorization-tool and/or OS resource usage. Our plan is at least verifying their usefulness and reduce the amount of false alarms (positives and/or negatives) by taking into account the combination of different tools and monitorization data collected at runtime. The minimum result, in this context, would be to be able to verify the (specific and sometimes complex) scaling services provided by public clouds.

References

1. Abbott, M.L., Fisher, M.T.: The Art of Scalability: Scalable Web Architecture, Processes, and Organizations for the Modern Enterprise, 2nd edn. Addison-Wesley Professional, Boston (2016). ISBN 0134032802
2. Anandhi, R., Chitra, K.: A challenge in improving the consistency of transactions in cloud databases – scalability. Int. J. Comput. Appl. 52(2), 12–14 (2012)
3. Davis, J., Daniels, R.: Effective DevOps: Building a Culture of Collaboration, Affinity, and Tooling at Scale. O'Reilly Media, Sebastopol (2016). ISBN 1491926309
4. Glitten. S.: Cloud vs. on-premises: finding the right balance. Computerworld, May 2017
5. Harchol-Balter, M.: Performance Modeling and Design of Computer Systems: Queueing Theory in Action. Cambridge University Press, Cambridge (2013). ISBN 1107027500
6. Hoffman, K.: Beyond the Twelve-Factor App. O'Reilly Media, Inc., Sebastopol (2016). ISBN 9781492042631
7. Jourdan, S., Pomes, P.: Infrastructure as Code (IAC) Cookbook Paperback, 17 February 2017

8. Michael, M., Moreira, J.E., Shiloach, D., Wisniewski, R.W.: Scale-up x scale-out: a case study using nutch/lucene. In: 2007 IEEE International Parallel and Distributed Processing Symposium (2007). https://doi.org/10.1109/ipdps.2007.370631

9. Morris, K.: Infrastructure as Code: Managing Servers in the Cloud. O'Reilly Media, Sebastopol (2016). ISBN 1491924357

10. Murphy, N.R., Beyer, B., Jones, C., Petoff, J.: Site Reliability Engineering: How Google Runs Production Systems. O'Reilly Media, Sebastopol (2016). ISBN 149192912X

11. Robinson, D.: Content Delivery Networks: Fundamentals, Design, and Evolution. Wiley, Hoboken (2017). ISBN 1119249872

12. Sill, A., Spillner, J. (eds.): 2018 IEEE/ACM 11th International Conference on Utility and Cloud Computing (UCC), Zurich, Switzerland. IEEE CPS, December 2018. ISBN 978-1-5386-5504-7

13. Hewlett Packard Enterprise, On-Premises Data Centers vs. Cloud Computing. https://www.hpe.com/us/en/what-is/on-premises-vs-cloud.html. Accessed 14 Mar 2019

14. "Proof of Concept on PHP Scaling", in Spanish, "Prueba de concepto sobre el escalado con PHP" (2019). https://github.com/chrodriguez/php-scale-probe

Intelligent Distributed System for Energy Efficient Control

Martín Pi Puig[1]([✉]) [iD], Juan Manuel Paniego[1] [iD], Santiago Medina[1] [iD],
Sebastián Rodríguez Eguren[1] [iD], Leandro Libutti[1] [iD],
Julieta Lanciotti[1], Joaquin De Antueno[1], Cesar Estrebou[1] [iD],
Franco Chichizola[1] [iD], and Laura De Giusti[1,2] [iD]

[1] Instituto de Investigación en Informática LIDI (III-LIDI),
Facultad de Informática, Universidad Nacional de La Plata (UNLP) - Comisión
de Investigaciones Científicas de la Provincia de Buenos Aires (CICPBA),
La Plata, Argentina
{mpipuig, jmpaniego, smedina, seguren, llibutti,
jlanciotti, jdeantueno, cesarest, francoch,
ldgiusti}@lidi.info.unlp.edu.ar
[2] Comisión de Investigaciones Científicas de la Provincia de Buenos Aires,
La Plata, Argentina

Abstract. In this work, we present an intelligent system developed for energy consumption distributed control and monitoring. It supports real time cloud-based data visualization of power profiles from different areas, so as to optimize overall power consumption.

The local intelligent processing unit (LIPU) that control the different environments is described. The communication network model that allows connecting multiple LIPUs to apply power consumption policies defined by the organization is analyzed, and the unit's capabilities in relation to cloud connectivity and real-time processing are considered through a theoretical scalability study.

Finally, we describe relevant implementation features in the context of "Facultad de Informática" of the "Universidad Nacional de La Plata" (Argentine).

Keywords: Energy consumption · Intelligent systems · Internet of things · Cloud computing · Optimization

1 Introduction

Nowadays there is a high energy demand in society for carrying out daily work or personal activities. In particular, the amount of energy consumed in public and private institutions increases each year due to the high number of electric devices used. On the other hand, electricity comes with a elevated cost and any unnecessary consumption involves additional costs that could have been avoided. A clear example of this are educational institutions such as schools and universities.

M. Naiouf et al. (Eds.): JCC&BD 2019, CCIS 1050, pp. 51–60, 2019.
https://doi.org/10.1007/978-3-030-27713-0_5

Additionally, there is a growing concern for environment preservation, which is taking governments throughout the globe in a search for solutions that can counter these situations [1].

According to the "Secretaría de Energía de la Nación" (SE), 87% of the primary energy consumed in Argentina comes from hydrocarbons. As a measure to reduce energy consumption, Decree 140/2007 proposes to reduce power consumption by 10% in public structures. Buildings account for around 40% of the final energy consumption and, as such, they offer a scenario with high potential to achieve significant reductions in energy consumption [2]. A highly energy-efficient building has a low environmental impact [3] while ensuring optimal interior conditions for the individuals, who spend more than 30% of their time in those spaces.

For these reasons, governments foster a set of policies and measures necessary to achieve energy savings in all sectors. An accurate knowledge of how much is actually consumed allows identifying unnecessary costs, optimizing the daily demand, balancing distributed loads and, as a consequence, decreasing energy consumption levels [4]. Then, power consumption can be optimized through the deployment of an intelligent system that collects information from building areas and act in consequence to save energy.

This system should consist of local intelligent processing units (LIPUs) that control the different environments and are connected to each other to obtain information for the entire building. The interconnecting network between these LIPUs should allow expanding the number of sectors that can be monitored within the building (or an area in it), and even consider groups of buildings. Toward this end, they could be connected to a server in the cloud that would be responsible for collecting the information from all separate buildings and areas.

Each LIPU needs a set or network of sensors that can detect events in the environment, such as the presence of people, temperature, devices connection status, and so forth. They should also have a set of actuators that allow controlling the state of the devices connected to the electrical power grid.

In this article, we explain the process used to develop the LIPUs as part of the Project "Unidad Inteligente para Control de Consumo Energético", approved by the Secretaría de Políticas Universitarias (SPU) as a technology transfer project within the "Universidades Agregando Valor 2017" program, and describe the communications network that allows linking several LIPUs belonging to the same building or area in to apply energy consumption policies defined by the organization. Finally, cloud connection capabilities are considered as a tool for monitoring geographically distributed locations and collecting the information generated in each of them for cloud-based tools analysis.

2 Intelligent Distributed System for Energy Consumption Control in Organizations: The Project

As already mentioned, public and private institutions consume large volumes of energy, so they need to implement measures and use intelligent systems to achieve significant savings. In this article, we analyze the case of the Universidad Nacional de

La Plata (UNLP), where the annual expense in energy consumption is 15% greater than the budget the University has assigned for non-salary expenses.

The UNLP consists of 17 schools with more than 100 buildings (including research, development and innovation units) and more than 1,200 physical spaces with diverse characteristics in relation to energy consumption requirements, such as:

- Classrooms: These are usually spacious rooms with plenty of lights, air-conditioning units, a projector, a computer and other elements. Generally, when the space is empty, all of these components can be turned off. On the other hand, they are laxly controlled and the electronic devices in them usually remain unnecessarily on.
- Offices: These are usually smaller spaces, with a few lights, an air-conditioning unit, several computers, a photocopier, and other elements. Generally, when the space is empty, all of these components can be turned off, except for the computers. They are more strictly controlled than classrooms, and usually electronic devices are not unnecessarily left on for long.
- Laboratories: These are usually small spaces, with a few lights, several air-conditioning units, a refrigerator/freezer, computers/servers, and other electronic equipment. In this case, none of these elements can be turned off, regardless of whether the space is empty or not.

Because of this, the UNLP is a good candidate to deploy and test an intelligent system to control power consumption. Taking into account the size and types of the spaces and the distribution of the buildings used by the UNLP, a layered system is appropriate. This requires considering at least 3 layers for the project (Fig. 1 shows a diagram of these 3 layers and how they relate to each other):

- The local intelligent processing units (LIPUs) and the possibilities for the sensor network to which they are connected.
- The network connecting these LIPUs to a local server (LS), by floor and/or building, so as to be able to implement policies based on physical spaces as well as on blocks of spaces (for instance, floors in a building).
- The connection to a cloud server (CS) and the analysis of cloud-based services to process energy consumption in real time for each unit, sector in a building, buildings, and as a whole for a physically distributed organization such as the UNLP.

In each monitorable environment (classroom, laboratory, office, etc.), a LIPU must be installed, as seen on the lower level of the diagram in Fig. 1. Since each environment has different dimensions, the number of intelligent devices that are part of a LIPU is variable. There is a master device that is responsible for controlling the environment by gathering information about other devices that are part of that LIPU to generate sector control guidelines.

On the second level shown in Fig. 1, there is a LS for each separate building, floor or area that will be monitored. This LS collects data from all LIPUs in each area to obtain current consumption statistics for each environment and take the necessary general control actions to minimize unnecessary expenses.

On the upper level of the diagram in Fig. 1, there is a CS that collects all data from each LS for the different separate buildings, floors or areas (which could be geographically separated from each other). This allows using cloud-based services to help process the energy consumed by each environment, building, or as a whole, and obtain detailed information about the overall consumption of the entire University.

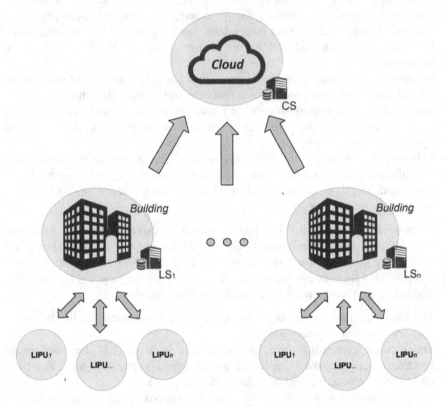

Fig. 1. Communication between the cloud and the geographically distributed buildings.

2.1 LIPUs Structure

A LIPU is composed by one or more hardware devices that offer a set of software functionalities, aimed at monitoring and controlling a number of electric components present in a given environment.

Since the scenarios have different dimensions, there are multiple devices interacting at the same time. There are two types of devices: masters and workers. Master devices are in charge of grouping, monitoring and controlling worker devices. Also, each master device can present functionalities that are similar to those offered by worker devices.

Figure 2 shows a generic representation of the internal structure of the devices that are part of a LIPU.

Even though the figure encompasses the distribution of hardware components in master and worker devices, each of these two types usually has different arrangements and functionalities. Generally, master control devices have a programming interface and an LCD display, combined with a real-time clock and, eventually, certain functionalities that are similar to the ones present in worker devices. On the other hand, workers are only responsible for processing and communicating data to the master. In other words, master units can offer the same functionalities that a worker device has, with the difference that they also allow human-machine interaction. This interface allows configuring the environment to be controlled.

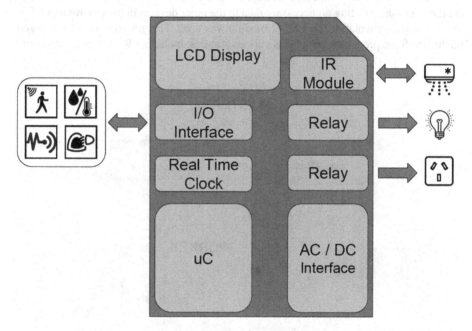

Fig. 2. Hardware diagram of LIPU components.

Even though the figure encompasses the distribution of hardware components in master and worker devices, each of these two types usually has different arrangements and functionalities. Generally, master control devices have a programming interface and an LCD display, combined with a real-time clock and, eventually, certain functionalities that are similar to the ones present in worker devices. On the other hand, workers are only responsible for processing and communicating data to the master. In other words, master units can offer the same functionalities that a worker device has, with the difference that they also allow human-machine interaction. This interface allows configuring the environment to be controlled.

As it can be seen, each device has an input/output interface that allows obtaining data from the different sensors, as well as executing different actions on a given electric device. With regard to input data, these units can be connected to different types of sensors that detect different variables such as presence, temperature, humidity, power current, luminosity, and so forth. The devices, otherwise, help to control the different electrical components through two output interfaces – relays and the infrared wave emitter. The former are electromagnetic devices that act as switches on an independent circuit, allowing turning the connected component on and off, for example, lights, power outlets, etc. Conversely, the IR emitter enables the wireless transmission of an on/off command for sophisticated devices where, turning them off forcibly (through relays) would directly affect their life time. An example of this type of devices are air-conditioning units. It should be noted that this IR module also supports learning from different infrared signals.

Unit processing and control is centralized on a micro-controller. This chip is also in charge of communicating all necessary data to the other devices in the environment. For this initial implementation, an ESP12 module was used. This micro-controller is based on the ESP8266 processor, which has various features, including Wi-Fi communication.

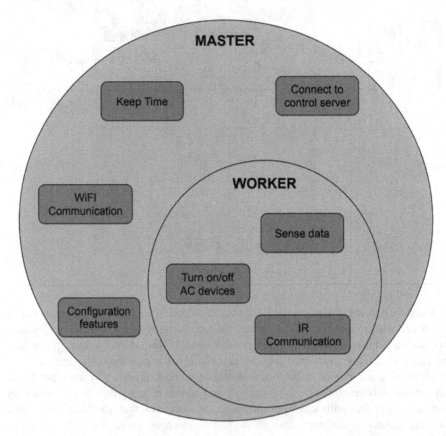

Fig. 3. Functionalities in a LIPU.

Master units can be programmed through a panel that has an LCD display, or using a Web application from a computer/tablet/mobile phone. Configurable functionalities include setting alarms, adding and removing sensors, defining events, adjusting the date, grouping relays, and so forth. Figure 3 shows a brief description of software features present in the system. The processes that master and worker devices can run are analyzed, and the tasks that can be carried out only by the master device are identified.

The devices that are part of a LIPU are connected by means of a local Wi-Fi network generated by the master device controlling the LIPU. This allows connecting a variable number of workers, each of them controlling different sensors/actuators. Because of this, each worker configuration is customized through the corresponding Wi-Fi network. Since control is centralized, the different workers monitor events in the environment and send the information to the master device for decision making.

This topology facilitates scalability to differently-sized scenarios just by increasing the number of sensors in existing devices, or by adding new worker devices.

It should be noted that LIPU devices consume a maximum of two to three orders of magnitude lower than the electrical components that could be found in a standard classroom. Table 1 details energy consumption per hour (in Wh) of both devices (master and worker) in LIPUs, and that of basic electrical components.

Using this information, the energy savings that the intelligent system identifies for each hour that should be turned off can be estimated. For instance, in a relatively small classroom, with 4 fluorescent tubes, a projector and one air-conditioning unit, the LIPU for the classroom requires two devices (one master and one worker). During the day, the LIPU will consume a total maximum of 144 Wh, which is compensated just by turning off the lights one hour during the day. If we take into account that there is also an air-conditioning unit and a projector, for each hour that the system establishes that all equipment should be turned off, more than 7500 Wh are saved (considering the consumption of these electronic devices).

Table 1. Energy consumption for various electronic devices.

	LIPU	Fluorescent tube	Projector	Air-conditioning unit
Energy consumption [Wh]	3	80	216	7300

2.2 Local Server (LS) in Each Building

One of the most common energy-related issues in buildings is the inefficient use of resources such as lights and electronic devices (air-conditioning units, fans, computers, projectors, and so forth). This can carry significant expenses and, eventually, result in various types of accidents due to device overheating and short-circuiting.

As a way to mitigate this issue, the LIPUs in a building or separate area can be grouped together so as to globally monitor and control all environments. To achieve this, there must be an existing wireless network covering the area delimited by the target scenarios, or a new one must be implemented. Being connected like this, the various master devices in the different LIPUs communicate the environment information and receive control parameters to act on it. The local server that centralizes control activities is implemented using a low-cost Raspberry Pi computer.

Building configuration is done through a Web system that integrates all environments and provides a simple interface for both global and specific customization for each environment. The system is coded using Python and the web framework Flask.

Toward this end, the system helps implement various control policies by allowing manual and automated settings. In the case of manual ones, they provide instant control over an environment, for example, to turn an electronic device on or off. On the other hand, it allows setting alarms (day/time to turn on/off a given electronic device), both sporadic and recurrent (for instance, weekly) for an automated control. Thus, each LIPU can act based on the information received from its connected sensors (mainly movement), or just by using a temporal restriction (alarms set).

Even though the system can be used for an entire building, it also provides a simple method to create a hierarchy of the structure to allow a layered approach to monitoring and configuration.

2.3   Cloud Server (CS)

The University needs to have centralized information about power consumption in its different buildings (or areas) to come up with suitable policies to optimize consumption. With the layers described in Sects. 2.1 and 2.2, this information is distributed among the different LSs, in detail. Therefore, a third layer is required for a CS to centralize this information, and a set of Amazon services to reduce information transfer from LSs to the CS and then process the information.

Service Architecture. A services architecture based on Amazon Web Service (AWS) public cloud is defined as part of the network of LIPUs and the LS in each building to help collect and process data, mainly focusing on power consumption measurements in each building. The AWS selected services are:

- *AWS IoT.* This service allows creating virtual devices on the cloud that represent the physical devices in the network, allowing two-way communication to collect data from the physical environment and send commands from the cloud.
  It allows working with different protocols, including HTTP and MQTT; as well as all necessary tools to define various security levels in the connection with each virtual device on the cloud [5].
  In our case, this service is used to represent on a virtual device each of the buildings where power consumption is being centrally measured.
- *AWS GreenGrass.* AWS GreenGrass allows taking the cloud environment to edge devices so that they can act locally with the data generated by the network of sensors to which they belong.
  This service is mainly used to pre-process the information generated by each of the components in the network, optimizing the number of packages that are sent to the cloud. This is particularly important, since Amazon services have a cost based on the flow of information; thus, when the number of packages sent is reduced by filtering out unnecessary data, general communication costs with the architecture will be lower as well [6].
  Specifically for our architecture, this service runs (on local servers) different algorithms to carry out the necessary operations to adapt the information that is

uploaded to the cloud. Then, a secure connection is established with AWS IoT to transmit the information.

- *AWS Lambda.* This Amazon tool allows executing code on demand. An algorithm is developed to work on a set of specific data and, based on predefined events, the code to process it is triggered.
  AWS Lambda can be configured and used both on the cloud and on the local server within AWS GreenGrass core [7].
- In our case, it is used to filter data and calculate some additional data based on the information collected by each LIPU in the network for the building. This algorithm is run on the LS before establishing a connection to the CS.

Experiments and Results. The experiments carried out include the deployment of a LS on a Raspberry Pi 3 with Raspbian as operating system and running AWS Green-Grass and a local instance of AWS Lambda with several preset functions. On the other hand, a virtual device is defined within AWS IoT in the cloud to receive the data processed from the network of LIPUs synchronized with the GreenGrass core in the LS.

The first test consisted in creating two devices that are connected to GreenGrass, to help them communicate through the central server. To do this, we employ MQTT protocol with certificates for a secure connection.

Then, the test was escalated by applying a Lambda function that responds to received message events for each node and creates packages with both messages and then synchronizes from GreenGrass to the virtual device on the cloud.

These two basic scenarios were successful, and we proceeded then to create a service architecture tailored to the needs of the different networks of LIPUs deployed in different buildings.

After this, with the purpose of creating a real test scenario, two Lambda functions were developed to control two different communication capabilities – one of these functions is triggered whenever a LIPU sends a message to its LS, and the other is run once every hour (this period of time can be configured) to synchronize data with the virtual device representing the building on the cloud.

Two alternatives were considered for this experiment:

- Each LIPU sends a message on a regular basis with power consumption information to the central controller where GreenGrass is being run (on the LS for the corresponding building).
- GreenGrass synchronizes these data by sending a message, also on a regular basis, with the global average of the consumption for the building being monitored.

It should be noted that GreenGrass service period is considerably longer than that of local nodes.

This experiment was useful to establish various guidelines for the local message flow handling based on the type of data used by the system. Also, certificates were used to provide secure communications, both over the local Wi-Fi as well as over the connection with AWS IoT.

3   Conclusions and Future Work

The architecture and corresponding software for an intelligent processing unit to control a network of sensors have been developed. This process is oriented to a type of physical spaces and equipment typical of University environments (classrooms, laboratories). These low-cost units help achieve significant savings in energy consumption in this type of environments and, since they can be programmed, they can be tailored to different spaces and types of sensors.

A distributed, layered architecture has been defined that allows LIPUs from different floors or buildings to connect to a network and through it be linked in real time to the cloud. This configuration allows scaling the solutions to the entire University (in our case, UNLP) with the possibility of monitoring critical aspects related to energy consumption in each building.

Cloud services have been utilized and communication and response time tests have been carried out to analyze project viability, obtaining good results. Based on our research and the tests carried out on Amazon's public cloud services, it can be states that the service architecture proposed in this article is suitable for the needs and characteristics of the general system. These services allow considering a scalable and elastic architecture, where it is easy to increase both the number of nodes per building as well as the overall number of buildings. Since the information is centralized using AWS IoT, only a Core GreenGrass deployment is required and then connecting each LIPU to its corresponding server using the same settings and, most importantly, the same security policies that protect the integrity of communications.

Future lines of work include a monetary study in relation to the serial manufacture of the intelligent units for power consumption control; scaling up the tests to the entire Facultad de Informática, which has 3 floors with 25 classrooms and laboratories, 3 server rooms and about 15 administrative offices; and expanding the tests from one building to connecting LIPUs in at least 3 schools of the UNLP located throughout La Plata.

References

1. Martínez, F.J.R., Gómez, E.V.: Eficiencia energética en edificios: certificación y auditorías energéticas. Thomson-Paraninfo, Madrid (2006)
2. Toranzo, E., Kuchen, E., Alonso-Frank, A.: Potenciales de eficiencia y confort para un mejor funcionamiento del edificio central de la universidad nacional de San Juan. AVERMA - Avances en Energías Renovables y Medio Ambiente **16**(1), 157–164 (2012)
3. Pérez-Lombard, L., Ortiz, J., Pout, C.: A review on buildings energy consumption information. Energy Build. **40**(3), 394–398 (2008)
4. Morán Álvarez, A.: Análisis y predicción de perfiles de consumo energético en edificios públicos mediante técnicas de minería de datos. Thesis of the University of Oviedo (2015)
5. Amazon Web Services IoT. https://aws.amazon.com/es/iot/. Accessed 25 Apr 2019
6. Amazon Web Services IoT Greengrass. https://aws.amazon.com/es/greengrass/. Accessed 28 Apr 2019
7. Amazon Web Services Lambda. https://aws.amazon.com/es/lambda/. Accessed 05 May 2019

Heap-Based Algorithms to Accelerate Fingerprint Matching on Parallel Platforms

Ricardo J. Barrientos[1]([✉])[ID], Ruber Hernández-García[1][ID], Kevin Ortega[2],
Emilio Luque[3][ID], and Daniel Peralta[4,5][ID]

[1] Laboratory of Technological Research in Pattern Recognition (LITRP),
Department of Computer Science and Industries, Faculty of Engineering Science,
Universidad Católica del Maule, Talca, Chile
{rbarrientos,rhernandez}@ucm.cl
[2] Kunert Business Software GmbH (KBS-Leipzig), Leipzig, Germany
kevin.ortega@kbs-leipzig.de
[3] Universitat Autònoma de Barcelona, Barcelona, Spain
emilio.luque@uab.es
[4] Data Mining and Modelling for Biomedicine Group, VIB Center for Inflammation
Research, Ghent, Belgium
daniel.peralta@irc.vib-ugent.be
[5] Department of Applied Mathematics, Computer Science and Statistics,
Ghent University, Ghent, Belgium

Abstract. Nowadays, fingerprint is the most used biometric trait for
individuals identification. In this area, the state-of-the-art algorithms
are very accurate, but when the database contains millions of identi-
ties, an acceleration of the algorithm is required. From these algorithms,
Minutia Cylinder-Code (MCC) stands out for its good results in terms
of accuracy, however its efficiency in computational time is not high. In
this work, we propose to use two different parallel platforms to accelerate
fingerprint matching process by using MCC: (1) a multi-core server, and
(2) a Xeon Phi coprocessor. Our proposal is based on heaps as auxiliary
structure to process the global similarity of MCC. As heap-based algorithms are exhaustive (all the elements are accessed), we also explored
the use an indexing algorithm to avoid comparing the query against all
the fingerprints of the database. Experimental results show an improvement up to 97.15x of speed-up, which is competitive compared to other
state-of-the-art algorithms in GPU and FPGA. To the best of our knowledge, this is the first work for fingerprint identification using a Xeon Phi
coprocessor.

Keywords: Coprocessors · Xeon Phi · MCC · Fingerprint

1 Introduction

Fingerprint identification is the most used biometric method to automatically
recognize the identity of a person [17], thanks to its usability and reliability

© Springer Nature Switzerland AG 2019
M. Naiouf et al. (Eds.): JCC&BD 2019, CCIS 1050, pp. 61–72, 2019.
https://doi.org/10.1007/978-3-030-27713-0_6

[23]. Fingerprints are the most studied biometric trait [11], and different algorithms have been proposed since 1975 to deal with them in their acquisition [20], processing [5], classification [11], and matching [6].

A fingerprint consists of a set of curves or lines, known as ridges. A ridge is defined as a single curved segment, and a valley is the region between two adjacent ridges. The discontinuities in the ridges, such as terminations and bifurcations, are called minutiae (see Fig. 1(a)). Formally, minutiae are points typically represented as a triplet (x, y, θ), where x and y represent the point coordinates and θ is the ridge direction at that point. These unique features are mathematically represented as a biometric template (also called template), which is stored in a biometric database. These templates are used in different ways for matching purposes. Although there are several fingerprint matching algorithms, the most common approaches are based on minutiae [11]. The objective of a minutiae-based algorithm is to find the maximum quantity of matching between pairs of two fingerprints.

Minutia Cylinder-Code (MCC) [6] is an accurate algorithm for fingerprint identification, which takes 45 ms for one comparison between two fingerprints. Thus, it implies 45 s for a database with 1000 fingerprints, which is a considerable time, especially when the database reaches the order of tens thousands or more fingerprints. There are two usual methods to decrease the execution time in this case, which are (1) to avoid comparing all fingerprint pairs by using classification methods [8,16] or indexing algorithms [4,7], and (2) to accelerate the matching processing by using parallel computing [21,27].

In this work, we explored to use both methods aiming to accelerate fingerprint matching process by using MCC. For this purpose, we employ a Xeon Phi coprocessor, which is one of the most promising alternatives for algorithms acceleration in the current technological context [29]. Besides, our proposal is based on heaps as auxiliary structure to process the global similarity of MCC. As heap-based algorithms are exhaustive (all the elements are accessed), we also use an indexing algorithm to avoid comparing the query against all the fingerprints of the database. There are several indexing algorithms to accelerate searching process on metric spaces [9,10,26]. In our approach, we selected the List of Clusters index [9] because it has shown good properties in different parallel platforms previously [1,12,25].

Thus, we propose a parallel algorithm based on heaps using a Xeon Phi coprocessor with the aim to obtain a high speed-up over the sequential counterpart, and showing how suitable is this parallel platform for fingerprint identification. To the best of our knowledge, this is the first approach with this coprocessor to accelerate fingerprint matching.

2   Related Work

2.1   Minutia Cylinder-Code

The Minutia Cylinder-Code algorithm [6], is currently one of the best methods to represent fingerprints and to execute a match making process. The main idea of

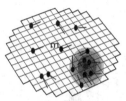

(a) Bifurcations and ridge endings are depicted in blue and red color, respectively. The orange circle locates the core of the fingerprint.

(b) Example of a neighborhood $N_{p_{i,j}^m}$: A cylinder section representing a neighborhood $N_{p_{i,j}^m}$ that is associated to a given minutia m in the center of this cylinder section. The darker section is the place where the highest point $p_{i,j}^m$ value lays (center of a cell), therefore it represents a higher contribution value C_m. The minutiae that are within the dark zone, are the neighborhood $N_{p_{i,j}^m}$.

Fig. 1. Minutiae and their representation by MCC.

the representation of the MCC is to generate a local structure for each minutia m in a given template T, where the structure is created by a spatial and directional relationship between the minutiae and their neighborhood, which is set according to a fixed radio [14]. Each local point is associated with a 3D structure, called cylinder. This cylinder is associated with each minutia m of a fingerprint (see Fig. 1(b)).

Figure 2(a) depicts the local structure of a cylinder. Each cylinder is centered on a minutia m that has a fixed radius and a height (from $-\pi$ to π) of 2π. In addition, all cylinders have the same size, because the fixed ratio. Each cylinder is discretized in $N_S \times N_S \times N_D$ cells. Each cell is a small cuboid with $\triangle_S \times \triangle_S$ base and $\triangle_D$ height, where $\triangle_S = \frac{2R}{N_S}$ and $\triangle_D = \frac{2\pi}{N_D}$ [6]. N_S is defined as the 2D space around a minutia m $N_S \times N_S$ and N_D represents the number of divisions that are applied to the height of 2π which represents the angular distance [14].

A numerical value is associated to each cell, also known as contribution $C_m(i, j, k)$, which is the sum of the contributions of each minutia mt belonging to the neighborhood $N_{p_{i,j}^m}$ at the point $p_{i,j}^m$ [22]. Based on the location and direction of each minutia in the neighborhood $N_{p_{i,j}^m}$, the values of both spatial and directional contributions are higher when: (1) the location is close to the point $p_{i,j}^m$, and (2) the direction is close to the angle set by $d\varphi_k$. By way of explanation [7], the value of a cell is a probability. This probability is higher when a minutia is close to a cell and its direction is similar to the value $d\varphi_k$ (see Fig. 2(b)).

A characteristic of MCC is that it can represent the value of each cell as a bit [6,7]. In this work, we propose algorithms to perform the matching of

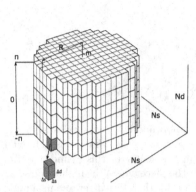

(a) Local structure of a cylinder associated to a minutia m in MCC.

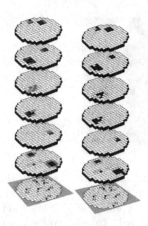

(b) Real (left) and binary (right) representation of the same cylinder.

Fig. 2. Local structure of a cylinder and its graphical representations.

fingerprints through the binary cylinders obtained from the MCC. The local similarity between two cylinders can be simply estimated by applying simple *bitwise* operations (*and, exclusive-or, population-count*) between the two corresponding binary vectors [6]. Thus, in order to compare two fingerprints, an overall score denoting their global similarity, has to be obtained from the local similarities. We used in the experiments as global similarity the Local Similarity Sort (LSS). This technique sorts the local similarity values and computes the average of the best n values, where the value of n is defined according to [6] by $n = min_{np} + \lfloor (Z(min\{n_A, n_B\}, \mu_P, \tau_P)) \cdot (max_{np} - min_{np}) \rfloor$, where n_A and n_B are the number of minutiae of the templates to be compared; $\lfloor \cdot \rfloor$ denotes the rounding operator; and Z is the function defined as $Z(v, \mu, \tau) = \dfrac{1}{1 + e^{-\tau(v-\mu)}}$. We used the same values proposed in [6] for $\mu_P = 20, \tau_P = 2/5, min_{np} = 4$, and $max_{np} = 12$.

2.2 Intel Xeon Phi Coprocessor

The Intel Xeon Phi coprocessor [28,29] consists of 61 to 72 cores connected by a high performance on-die bidirectional interconnect. The coprocessor runs a Linux operating system and supports all main Intel development tools like C/C++, Fortran, MPI and OpenMP. The coprocessor is connected to an Intel Xeon processor (the host) via the PCI Express (PICe) bus. It is mainly composed of cores, which have one Vector Processing Unit (VPU), and one L1 and L2 cache per core. In the VPU each operation can be a fused multiply-add giving 32 single-precision or 16 double-precision floating-point operations per cycle. This architecture has a complex vector unit. However, empirical studies show that the efficient exploitation of the vector unit is crucial to achieve a significant performance improvement [28]. A limitation of using a Xeon Phi, such as in other coprocessors, is to deal with the transfers to/from the CPU.

2.3 Approaches Based on Parallel Computing

The algorithm MCC has shown very high performance in terms of accuracy, but it requires a high computation cost. In the Parallel Computing area, the use of coprocessors to accelerate processing is being exploited, and currently there are three main coprocessors used for this purpose: GPU, FPGA, and Intel Xeon Phi.

In the present work, beside proposing and implementing a multi-core algorithm, we also developed a Xeon Phi algorithm, which is to the best of our knowledge the first algorithm with this coprocessor to accelerate fingerprint matching. We compare our algorithms to previous state-of-the-art approaches that use different parallel platforms with the MCC algorithm [15,18,22].

Gutierrez et al. [15] implement an algorithm based on GPU for the MCC. They use two different models of GPU in their experiments. They implement an exhaustive algorithm where all the fingerprints are accessed by each query. They achieve a speed-up of 29.0x with the LSS similarity and $N_s = 8$. Lindoso et al. [22] propose an algorithm based on a FPGA for the fingerprint matching, achieving a speed-up of 23.7x. Other work based on FPGA is proposed by Jiang and Crookes [18], which is to the best of our knowledge the FPGA-based algorithm with highest performance, achieving a speed-up of 46.5x.

3   Parallel Computing Algorithms

In this section we show our proposed algorithms for fingerprint identification on a multi-core server (Sect. 3.1) and a Xeon Phi coprocessor (Sect. 3.2). We also explored the use of an indexing method to accelerate the search by discarding elements with a pre-processing of the database (Sect. 4.1).

3.1   Multi-core Algorithm

We take as input parameter the cylinders of each fingerprint, where each cylinder is an array of bits. This algorithm is illustrated in Fig. 3, where we distributed the queries among the threads of the multi-core server. Each thread performs the similarity function against all the elements of the database according to the similarity of cylinders described in Sect. 2.1. For avoiding O.S. resource conflicts, we execute each thread in an exclusive core.

In our algorithm the data (database and queries) is store in two matrices, but for this we keep an index indicating where start and end the cylinder of each fingerprint. Thus, we used a heap [19] as auxiliary structure to keep the n_p highest similarities among cylinders. Each thread creates its own heap as a private variable. This is required for the global similarity LSS that we used as the global and final similarity.

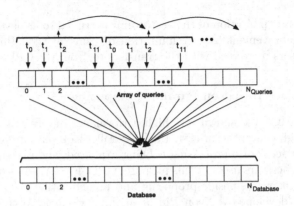

Fig. 3. Illustration of the muti-core algorithm.

3.2 Xeon Phi Algorithm

Given a fingerprint database of N_{DB} fingerprints. Let $Fingerprints$ be the matrix storing the cylinders of N_{DB} fingerprints in the database; $Queries$ be the matrix storing $num\_queries$ queries; and ID_thread be the current thread ID and $num\_threads$ the total number of threads. We propose an algorithm based on MCC for a Xeon Phi coprocessor, which is presented in Algorithm 1. The queries are accessed in a round robin manner (line 4 in Algorithm 1) by the thread of the Xeon Phi. Each thread calculates a local similarity between its query and the fingerprints in the database (line 14), and then the global similarity (line 26) with the values stored in its heap of size n, which is the overall similarity score between the two fingerprint templates to be compared.

This algorithm does not require any synchronization instructions. The only synchronization of the algorithm is carried out implicitly when the pragma offload region ends. It is noteworthy that this algorithm has focused on processing the queries in batches of size $num\_queries$.

4 Experimental Results

In our experiments, we used a multi-core server composed of two Haswell architecture Intel Xeon E5-2620v3 2.4 GHz, i.e. 12 hyperthreading cores with 32 GB

Table 1. Platforms description.

(a) Intel Xeon Phi details.

Coprocessor	Intel Xeon Phi 7120P
Cores	61 cores of 1.24 GHz / 4 threads per core
Memory	16GB of memory (bandwidth 352 GB/s)
Cache L1	3.8MB L1 (64KB L1 per core)
Cache L2	30.5MB L2 (512KB L2 per core)
Compiler	icc version 15.0.2, flags: -O3 -openmp

(b) Platform for sequential and multi-core versions

Processor	2xIntel Xeon E5-2660v3, 12 cores with HT 15MB Cache, 2.40 GHz, Haswell
Memory	32 GB, 59.7 GB/s
Operative System	GNU Debian System Linux kernel 3.10.0-229.4.2.el7.x86_64
Compiler	icc version 15.0.2, flags: -O3

Algorithm 1. Algorithm in Xeon Phi based on the MCC.

```
 1: #pragma offload in(Fingeprints) in(Queries) in(Heaps)
 2: {
 3:    #pragma omp parallel
 4:    for (i = ID_thread; i < num_queries; i += num_threads)
 5:      for (j = 0; j < N_DB; j++)
 6:      {
 7:          first_cyl = start_cylinders(Fingeprints, j);
 8:          size_cyl = size_cylinders(Fingerprints, j);
 9:          first_cyl_Q = start_cylinders_query[i];
10:          size_cyl_Q = size_cylinders_query[i];
11:          for (k=first_cyl_Q; k <first_cyl_Q+size_cyl_Q; k++)
12:            for (l=first_cyl; l <first_cyl+size_cyl; l++)
13:            {
14:                similarity = 1 − (‖Fingerprints[l]XORQueries[k]‖)/(‖Fingerprints[l]‖+‖Queries[k]‖)
15:                if (Heap[ID_thread].size() >= n)
16:                {
17:                    if (Heap[ID_thread].top() < similarity)
18:                    {
19:                      Heap[ID_thread].pop();
20:                      Heap[ID_thread].push(similarity);
21:                    }
22:                    else
23:                      Heap[ID_thread].push(similarity);
24:                }
25:            }
26:            global_similarity = LSS(Heap[ID_thread]);
27:      }
28: }
```

in main memory. The coprocessor used in the experiments was an Intel Xeon Phi 7120 with 61 cores, supporting up to 4 threads per core, and 16 GB of memory. We used all cores in our Xeon Phi algorithm, and 4 threads per core. More details are shown in Table 1.

We used the NIST (National Institute of Standards and Technology) [30] as database, which contains 27,000 pairs of segmented 8-bit gray scale fingerprint images. We used the first 27,000 impressions as database and the 27,000 s impressions as queries. The average number of minutiae is 206.9 with a maximum of 610.

Figure 4 shows performance measures between the sequential and multi-core method. Figure 4(a) shows the speed-up of the multi-core algorithm, and Fig. 4(b) shows the quantity of cylinders comparison operations performed per second. Both experiments in Fig. 4 execute one thread in one exclusive core, and for the label 24(HT) we executed two threads per core using the hyperthreading property of the processors. We observe a good scaling behavior when a thread is executed in an exclusive core, but also with 24 threads (taking into account the shared resources for hyperthreading cores).

Figure 5 shows the speed-up of the sequential and multi-core algorithms with 12 and 24 HT threads against the algorithm in Xeon Phi. It should be highlighted each core in Xeon Phi supports up to 4 threads executed simultaneously. Our

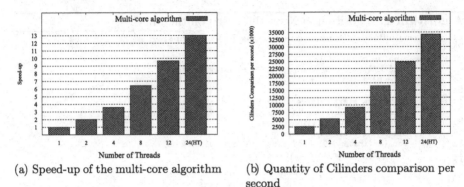

(a) Speed-up of the multi-core algorithm

(b) Quantity of Cilinders comparison per second

Fig. 4. Performance measures between the Sequential and multi-core version. HT = Hyperthreading.

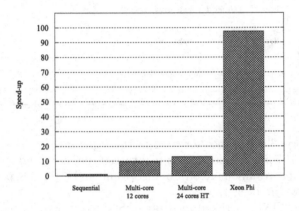

Fig. 5. Speed-up between of the multi-core and Xeon Phi version. Values obtained over the sequential algorithm. HT = Hyperthreading.

algorithm in Xeon Phi reaches 97.51x of speed-up because of the good use of cache when threads executed in the same core access to the same data [28]. This behavior occurs when threads processing different queries must access to the same elements of the database.

There are different previous works covering the fingerprint identification in the coprocessors GPU and FPGA, which were described in Sect. 2.3. We compare our results to these previous state-of-the-art algorithms in the Table 2. It should be noticed that FPG-based approaches use different databases and they achieve a speed-up up to 46.5x. Our algorithm in Xeon Phi achieves a competitive speed-up, showing that a Xeon Phi coprocessor can be used for fingerprint matching and specifically for the MCC algorithm.

Table 2. Comparison with state-of-the-art algorithms on different parallel platforms.

Algorithm	Parallel platform	Database	Speed-up
Multi-core algorithm	24 (HT) threads	NIST Special Database 14	12.99x
Xeon Phi algorithm	Xeon Phi 7120	NIST Special Database 14	97.51x
GPU algorithm with LSS in [15]	NVIDIA GeForce GTX 680	NIST Special Database 14	29.0x
GPU algorithm with LSS in [15]	NVIDIA Tesla M2090	NIST Special Database 14	24.9x
FPGA method in [18]	FPGA Xilinx Virtex-E	10,000 fingerprints randomly generated	46.5x
FPGA method in [22]	FPGA Xilinx Virtex-4 LX	56 fingerprints	23.7x

4.1   Indexing Algorithm for MCC

In recent years, in the area of similarity search, many indexing approaches have been proposed [10]. The objective of these indexes is to avoid distance calculations between the query and each of the elements of the database. This is performed by discarding elements using the triangle inequality property of a metric space. In this work, we implemented the List of Cluster (LC) [9] index to be used with the MCC algorithm, and trying of discarding fingerprints of the database.

We used an adaptation of the cylinder similarity function to be able of using the LC index, which requires that the data (X) and similarity function (d) be a metric space [31]. This means that d must hold the following properties on X: (1) strict positiveness $(d(x,y) > 0$ and if $d(x,y) = 0$ then $x = y)$, (2) symmetry $(d(x,y) = d(y,x))$, and the triangle inequality $(d(x,z) \leq d(x,y) + d(y,z))$.

We used the similarity function shown in [7], where given two fingerprints F_1, F_2, and S_1 S_2 be the corresponding sets of n-dimensional binary vectors, a similarity measure between F_1 and F_2 can be defined as follows:

$$sim(F_1, F_2) = \frac{\sum_{s \in S_1} max_{s_j \in S_2}\{H(s, s_j)\}}{|S_1|} \qquad (1)$$

where $|S_1|$ is the cardinality of set S_1, and H is a normalized similarity measure between two binary vectors, based on the Hamming distance (d_H), defined as follows:

$$H(a, b) = \left(1 - \frac{d_H(a, b)}{dim}\right) \qquad (2)$$

where dim is the dimension of the binary vectors, and p is a parameter controlling the shape of the similarity function (we set $p = 30$ for the experiments).

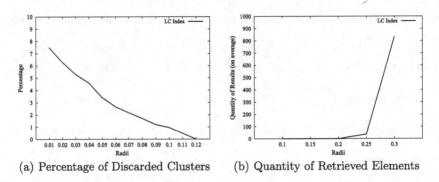

(a) Percentage of Discarded Clusters   (b) Quantity of Retrieved Elements

Fig. 6. Average values using the List of Cluster index.

We selected the LC index because: (1) it has been previously used with parallel platforms [2,13,24] (2) they hold their indexes in dense matrices which are very convenient data structures for mapping algorithms onto Xeon Phi. This index can be implemented dividing the space in two different ways: taking a fixed radius for each partition or using a fixed size. To ensure good load balance in a parallel platform, we consider partitions with a fixed size of K elements, thus the cluster radius rc with the center C is the maximum distance between its center and its Kth nearest element.

We did not achieve positive results using the LC index, obtaining execution times higher than the exhaustive approaches. This is because this index was not able to discard elements with the elements of the MCC algorithm. We can see in Fig. 6(a) the percentage (on average) of discarded elements when the radius of the query is increased, and Fig. 6(b) shows the number of retrieved elements when the radius of the query is increased. We observe that when the radius is large enough to retrieve elements, the LC is not able to discard any elements. This phenomenon has been observed before, and it is named as *curse of dimensionality* [3], where the intrinsic dimensionality [10] is high enough to avoid discarding, and the distance histogram of the elements is more concentrated when the dimension grows. Despite the fact that we did not achieve a good result in performance, we decided to add this section to conclude that a metric index like the LC is not efficient with this algorithm, because of the nature of the data.

5   Conclusions

In this work, we have proposed and implemented algorithms for fingerprint identification on two different kinds of parallel platforms: a multi-core server and a Xeon Phi coprocessor.

Our algorithms are based on heaps and exhaustive search, but we also explored an indexing approach using the List of Cluster index. This index did not reach a good performance, mainly because the nature of the vectors did not allow discarding of elements, which is the purpose of the index. This is because the intrinsic dimension of the space is high and it is affected by the well known curse of dimensionality.

Our algorithms achieve up to 97.51x of speed-up, outperforming previous state-of-the-art approaches in GPU and FPGA. Our results show that the computation need of the MCC algorithm can be covered and accelerated using a Xeon Phi. This coprocessor has the advantage of a reduced cost compared to a conventional multi-core server, and also a lower energy consumption. To the best of our knowledge, this is the first work for fingerprint identification using a Xeon Phi coprocessor.

Acknowledgement. This research was partially funded by Project CONICYT FONDEF/Cuarto Concurso IDeA en dos Etapas del Fondo de Fomento al Desarrollo Científico y Tecnológico, Programa IDeA, FONDEF/CONICYT 2017 ID17i10254. D. Peralta is a Postdoctoral Fellow of the Research Foundation of Flanders.

References

1. Barrientos, R.J., Gómez, J.I., Tenllado, C., Matias, M.P., Marin, M.: kNN query processing in metric spaces using GPUs. In: Jeannot, E., Namyst, R., Roman, J. (eds.) Euro-Par 2011. LNCS, vol. 6852, pp. 380–392. Springer, Heidelberg (2011). https://doi.org/10.1007/978-3-642-23400-2_35
2. Barrientos, R.J., Gómez, J.I., Tenllado, C., Matias, M.P., Marin, M.: Range query processing on single and multi GPU environments. Comput. Electr. Eng. **39**(8), 2656–2668 (2013)
3. Bellman, R.: Adaptive Control Processes: A Guided Tour. A Rand Corporation Research Study Series. Princeton University Press, Princeton (1961)
4. Bhanu, B., Tan, X.: A triplet based approach for indexing of fingerprint database for identification. In: Bigun, J., Smeraldi, F. (eds.) AVBPA 2001. LNCS, vol. 2091, pp. 205–210. Springer, Heidelberg (2001). https://doi.org/10.1007/3-540-45344-X_29
5. Cao, K., Liu, E., Jain, A.K.: Segmentation and enhancement of latent fingerprints: a coarse to fine ridgestructure dictionary. IEEE Trans. Pattern Anal. Mach. Intell. **36**(9), 1847–1859 (2014)
6. Cappelli, R., Ferrara, M., Maltoni, D.: Minutia cylinder-code: a new representation and matching technique for fingerprint recognition. IEEE Trans. Pattern Anal. Mach. Intell. **32**(12), 2128–2141 (2010)
7. Cappelli, R., Ferrara, M., Maltoni, D.: Fingerprint indexing based on minutia cylinder-code. IEEE Trans. Pattern Anal. Mach. Intell. **33**(5), 1051–1057 (2011)
8. Cappelli, R., Maio, D.: The state of the art in fingerprint classification. In: Ratha, N., Bolle, R. (eds.) Automatic Fingerprint Recognition Systems, pp. 183–205. Springer, New York (2004). https://doi.org/10.1007/0-387-21685-5_9
9. Chávez, E., Navarro, G.: A compact space decomposition for effective metric indexing. Pattern Recogn. Lett. **26**(9), 1363–1376 (2005)
10. Chávez, E., Navarro, G., Baeza-Yates, R., Marroquín, J.L.: Searching in metric spaces. ACM Comput. Surv. **33**(3), 273–321 (2001)
11. Galar, M., et al.: A survey of fingerprint classification part i: taxonomies on feature extraction methods and learning models. Knowl.-Based Syst. **81**, 76–97 (2015)
12. Gil-Costa, V., Barrientos, R.J., Marin, M., Bonacic, C.: Scheduling metric-space queries processing on multi-core processors. In: 18th Euromicro Conference on Parallel, Distributed and Network-based Processing (PDP 2010), pp. 187–194. IEEE Computer Society, Pisa (2010)

13. Gil-Costa, V., Marin, M.: Load balancing query processing in metric-space similarity search. In: 12th IEEE/ACM International Symposium on Cluster, Cloud and Grid Computing (CCGRID 2012), pp. 368–375. IEEE, Ottawa (2012)

14. Gutiérrez, P.D., Lastra, M., Herrera, F., Benítez, J.M.: A high performance fingerprint matching system for large databases based on GPU. IEEE Trans. Inf. Forensics Secur. 9(1), 62–71 (2014)

15. Gutierrez, P.D., Lastra, M., Herrera, F., Benitez, J.M.: A high performance fingerprint matching system for large databases based on GPU. IEEE Trans. Inf. Forensics Secur. 9(1), 62–71 (2014)

16. Hong, J.H., Min, J.K., Cho, U.K., Cho, S.B.: Fingerprint classification using one-vs-all support vector machines dynamically ordered with Naï ve Bayes classifiers. Pattern Recogn. 41(2), 662–671 (2008)

17. Jain, A., Flynn, P., Ross, A.A.: Handbook of Biometrics. Springer, New York (2007). https://doi.org/10.1007/978-0-387-71041-9

18. Jiang, R.M., Crookes, D.: FPGA-based minutia matching for biometric fingerprint image database retrieval. J. Real-Time Image Proc. 3(3), 177–182 (2008)

19. Knuth, D.E.: The Art of Computer Programming, vol. 3. Addison-Wesley, Boston (1973)

20. Kumar, A., Kwong, C.: Towards contactless, low-cost and accurate 3D fingerprint identification. In: Proceedings of the IEEE Conference on Computer Vision and Pattern Recognition, pp. 3438–3443 (2013)

21. Le, H.H., Nguyen, N.H., Nguyen, T.T.: Exploiting GPU for large scale fingerprint identification. In: Nguyen, N.T., Trawiński, B., Fujita, H., Hong, T.-P. (eds.) ACIIDS 2016. LNCS (LNAI), vol. 9621, pp. 688–697. Springer, Heidelberg (2016). https://doi.org/10.1007/978-3-662-49381-6_66

22. Lindoso, A., Entrena, L., Izquierdo, J.: FPGA-based acceleration of fingerprint minutiae matching. In: 2007 3rd Southern Conference on Programmable Logic, pp. 81–86 (2007)

23. Maltoni, D., Maio, D., Jain, A., Prabhakar, S.: Handbook of Fingerprint Recognition. Springer, London (2009). https://doi.org/10.1007/978-1-84882-254-2

24. Marin, M., Gil-Costa, V.: Approximate distributed metric-space search. In: ACM Workshop on Large-Scale and Distributed Information Retrieval (LSDS-IR 2011), Glasgow, UK (2011)

25. Marin, M., Gil-Costa, V., Bonacic, C., Baeza-Yates, R., Scherson, I.D.: Sync/async parallel search for the efficient design and construction of web search engines. Parallel Comput. 36(4), 153–168 (2010)

26. Navarro, G., Uribe-Paredes, R.: Fully dynamic metric access methods based on hyperplane partitioning. Inf. Syst. 36(4), 734–747 (2011)

27. Peralta, D., Triguero, I., Sanchez-Reillo, R., Herrera, F., Benítez, J.M.: Fast fingerprint identification for large databases. Pattern Recogn. 47(2), 588–602 (2014)

28. Partnership for Advanced Computing in Europe (PRACE): Best Practice Guide - Intel Xeon Phi

29. Wang, E., et al.: High-Performance Computing on the Intel® Xeon Phi™. Springer, Cham (2014). https://doi.org/10.1007/978-3-319-06486-4

30. Watson, C.I.: NIST Special Database 14. Fingerprint Database, US National Institute of Standards and Technology (1993)

31. Zezula, P., Amato, G., Dohnal, V., Batko, M.: Similarity Search: The Metric Space Approach. Advances in Database Systems, vol. 32. Springer, New York (2006). https://doi.org/10.1007/0-387-29151-2

Big Data and Data Intelligence

An Analysis of Local and Global Solutions to Address Big Data Imbalanced Classification: A Case Study with SMOTE Preprocessing

María José Basgall[1,2,3](✉) [ID], Waldo Hasperué[2][ID], Marcelo Naiouf[2][ID],
Alberto Fernández[4][ID], and Francisco Herrera[4][ID]

[1] UNLP, CONICET, III-LIDI, La Plata, Argentina
mjbasgall@lidi.info.unlp.edu.ar
[2] Instituto de Investigación en Informática (III-LIDI),
CIC-PBA Facultad de Informática - Universidad Nacional de La Plata,
La Plata, Argentina
[3] University of Granada, Granada, Spain
[4] DaSCI Andalusian Institute of Data Science and Computational Intelligence,
University of Granada, Granada, Spain

Abstract. Addressing the huge amount of data continuously generated is an important challenge in the Machine Learning field. The need to adapt the traditional techniques or create new ones is evident. To do so, distributed technologies have to be used to deal with the significant scalability constraints due to the Big Data context.

In many Big Data applications for classification, there are some classes that are highly underrepresented, leading to what is known as the imbalanced classification problem. In this scenario, learning algorithms are often biased towards the majority classes, treating minority ones as outliers or noise.

Consequently, preprocessing techniques to balance the class distribution were developed. This can be achieved by suppressing majority instances (undersampling) or by creating minority examples (oversampling). Regarding the oversampling methods, one of the most widespread is the SMOTE algorithm, which creates artificial examples according to the neighborhood of each minority class instance.

In this work, our objective is to analyze the SMOTE behavior in Big Data as a function of some key aspects such as the oversampling degree, the neighborhood value and, specially, the type of distributed design (local vs. global).

Keywords: Big Data · Imbalanced classification ·
Preprocessing techniques · SMOTE · Scalability

© Springer Nature Switzerland AG 2019
M. Naiouf et al. (Eds.): JCC&BD 2019, CCIS 1050, pp. 75–85, 2019.
https://doi.org/10.1007/978-3-030-27713-0_7

1   Introduction

Currently, Data Science has an essential role on analyzing the enormous amount of data being generated in every moment. This is known as Big Data, and the more volume of available information, the more knowledge could be discovered [1].

However, it is known that the "small data" or standard size problems implementations are not directly applicable to Big Data due to the scalability constraints [2]. For this reason, the traditional techniques have to be adapted to the "divide-and-conquer" approach proposed by "the facto" MapReduce [3] framework for Big Data. In this direction, two alternatives are known, the local and the global design approaches [4]. The former works with each data partition separately and the results of each Map process is put together on a single Reduce process. And the latter, generates global results by distributing data and models across the Map processes. This is considered as an exact model because the final results are obtained through a more complete insight of the data.

It has to be considered that this huge amount of data does not imply all of it will be useful. Indeed, most of the times, a subset of the data will be the real source of the knowledge discovery process. As it happens with "small data", the results of this process are directly related to the quality of the data used. Thus, to obtain high quality data (also known as Smart Data [5]), preprocessing techniques have to be applied.

In order to study the data quality in a Big Data context, the focus is set on a common situation when a classification problem is faced: imbalanced (or uneven) data distribution. Imbalanced data classification is a meaningful topic due to the large amount of real problems in which the key concept is represented by the minority class (e.g., medical diagnosis of rare diseases).

In this research area, the existent methods for balancing data are undersampling and oversampling. They work eliminating or creating, majority or minority class instances, respectively. With respect to the oversampling methods, the SMOTE ("Synthetic Minority Oversampling TEchnique") [6,7] algorithm is one of the most widespread. SMOTE creates artificial examples by interpolation according to the neighborhood of each minority class instance.

In this work, an analysis of the current preprocessing solutions for imbalanced Big Data behavior is carried out. A performance comparison of the solutions related to parameters of interest, such as the number of partitions and oversampling final ratio is shown. In addition, and regarding the SMOTE algorithm, the focus is on contrasting the behavior between the global and the local scheme implementations. The main objective is to analyze the obtained results to determine the dependency of the imbalance preprocessing or the data intrinsic quality. Another aspect to evaluate is if their performance behave as in traditional datasets context.

The article continues organized as follows. In Sect. 2, the current solutions for imbalanced Big Data classification are described. Section 3 details the experimental environment used in this work. Then, in Sect. 4, the comparative results are shown. Finally, in Sect. 5, conclusions and future works are described.

2   Big Data and the Imbalanced Classification Problem

In this section, a brief introduction to the most used Big Data frameworks is presented in Sect. 2.1. Furthermore, a quick review about imbalanced classification and a description of its methods for Big Data are depicted in Sect. 2.2.

2.1   Big Data Technologies

Due to Big Data, new technologies appeared in order to cope with it. Among them, in 2003 and developed by Google, the most significant was born: MapReduce [3]. This framework was design based on a "divide-and-conquer" scheme in order to process Big Data on a cluster using parallel and distributed implementations. MapReduce model presents two stages called Map and Reduce. The former receives data and performs operations in order to transform them. The latter process the results of the previous phase to summarize them. This model works with key-value pairs. In order to process them in parallel, all the pairs of the same key are distributed to the same node.

The most popular open-source frameworks based on MapReduce model programming are Apache Hadoop [8] and Apache Spark [9,10]. The main difference between them is that Hadoop performs an intensive disk usage, and Spark an intensive memory usage. This generates that Spark outperforms Hadoop. Also Spark provides integration with many libraries such as MLlib [11] (the Machine Learning library), Spark Streaming [12] (to work with streams of data), among others. These are some of the reasons which make Spark the current widespread Big Data framework.

In Sect. 1 two design methods related to the use of data and models distribution were depicted: the local and the global [4]. Depending on which model is applied, the results of the developed algorithm will be approximated or exact.

2.2   Imbalanced Classification in Big Data

In a classification task, poor quality or not optimal data, will imply that the results will neither be. A previous process of adaptation has to be applied on data to carry out a good learning. A common scenario to apply classifiers is when the dataset class distribution is imbalanced. The simplest preprocessing techniques to achieve a balanced dataset are ROS (Random Over Sampling) and RUS (Random Under Sampling) algorithms [13]. On one hand, ROS works replicating the minority instances in a random way in order to achieve the desired ratio balance between both classes. On the other hand, RUS creates a balanced dataset by random deletion of the majority instances.

As was introduced in Sect. 1, the SMOTE algorithm [6,7] is one of the most applied on the oversampling preprocessing cases. SMOTE works over the minority class instances (also known as positive instances) by calculating the k nearest neighbors (kNN) of each of them. This technique creates a balanced dataset interpolating each of the positive examples with its neighbors as can be seen in Fig. 1.

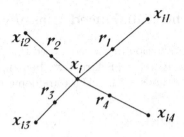

Fig. 1. Interpolation between a minority instance and its k nearest neighbors ($k = 4$).

In the MapReduce approach, the data partitioning task may lead to lack of data when processing local models. Furthermore, it could also cause "small-disjuncts" [14]. This extreme lower number of minority instances in each Map gives an incomplete representation of the dataset information. In particular, regarding the real neighborhood of each example. "Small-disjuncts" are tiny groups of very local data and with low density, surrounded by the majority class instances. Minority instances created from them may enter in the exclusive zones of the majority class inducing noise or an over-generalization.

To the best of our knowledge, no other solutions than the ROS, RUS and the SMOTE are currently available as preprocessing techniques for imbalanced Big Data problems.

In [15], authors present a work in which these methods have been adapted to the MapReduce programming style, making them suitable for Big Data. They are called RUS-BigData and ROS-BigData respectively, and both of them have been designed using the local approach. Each Map process adjusts the class distribution for the data that belongs to it by performing the under or the over sampling. Thus, these are solutions independent of the partitions number. Then, to obtain the balanced dataset, a single Reduce process collects the results generated by each Map.

Regarding the SMOTE algorithm, in [16] a global SMOTE fully scalable solution was described, called SMOTE-BD. In order to cope with the potential data partitioning problems, the whole neighborhood of each minority class instance is taken into account. That is achieved by the use of scalable data structures. The source code of SMOTE-BD can be found as a package in the Spark-packages repository [17]. The k nearest neighbors (kNN) calculation was based on [18].

Furthermore, a local SMOTE version called SMOTE-MR is available in [19]. The k nearest neighbors for each minority instance are obtained from data belonging to the same instance's partition. Working independently on each Map, gives approximated final results. This is the reason why methods are called global (or exact) and local (or approximated) as mentioned before.

Some aspects to remark are that the exact approach requires more effort in the solution development than the approximated one (which is straightforward to the MapReduce programming model). And the main advantage of the exact design is the learning of more robust models due to the capacity of sharing data and models [4,20].

All of the mentioned preprocessing techniques in this section were developed using Apache Spark.

3 Experimental Framework

In this section, the experimental environment is detailed. First, in Sect. 3.1 the datasets and the algorithms parameters configuration used in the tests are enumerated. Then, the classifier and the evaluation metrics are described in Sect. 3.2. Finally, in Sect. 3.3 the infrastructure used for the experiments is mentioned.

3.1 Datasets and Algorithms Parameters

In order to compare the performance of the four preprocessing algorithms for Big Data, three imbalanced datasets were selected from the UCI Machine Learning repository [21], each one with a very different imbalanced ratio.

Table 1 shows the datasets summary, where the number of examples ($\#Ex.$), number of attributes ($\#Atts.$), number of instances for each class ($\#(maj; min)$), class distribution ($\%(maj; min)$) and imbalance ratio (IR) are included.

Table 1. Datasets summary

Datasets	#Ex.	#Atts.	#(maj; min)	%(maj; min)	IR
covtype7	464,677	54	(448,421; 16,256)	(96.5; 3.5)	27.58
higgs	4,954,754	28	(4,663,298; 291,456)	(94.12; 5.88)	16
susy	2,212,186	18	(2,169,299; 542,435)	(80; 20)	4

The parameters used for the methods according to their authors' specifications are shown in Table 2. As can be seen, a division was made in order to show the parameters in common for all the algorithms and the specific for the SMOTEs.

The percentage of oversampling (% oversampling) represents the final desired distribution between classes, that is, the final ratio. For instance, if % oversampling = 100, both classes will have the same quantity of instances, in other words, a 1:1 ratio; and if % oversampling = 150, the minority class will end up with a 50% more of instances than the majority class, this mean, a 1.5:1 ratio, and so on.

There are several studies where the percentage of oversampling had influenced significantly over the results [22] as they increase. In consequence, three values were selected for this parameter with the purpose to test each of them.

The number of partitions (# partitions) sets the amount of Map process to be used. That means the number in which input data will be split.

Regarding both SMOTE versions, as they calculate the k Nearest Neighbors of each minority instance, different values of the k parameter have been proposed. The euclidean distance function was used.

Table 2. Algorithms and parameters

Algorithms	Parameters	Values
All	% oversampling	100/150/200
	# partitions	32/64/128
SMOTEs	k Nearest Neighbors	3/5/7
	Distance function	Euclidean

3.2   Classifier and Evaluation Metrics

The behavior of the resultant preprocessed datasets was tested using the Decision Trees classifier (DT), implemented in the Spark's MLlib library [11]. The Apache Spark and MLlib version used for this work was the 2.2.0.

Table 3 shows a confusion matrix for a binary problem from which the classification quality metrics are obtained. This matrix organizes the samples of each class according to their correct or incorrect identification. Thus, the prediction quality for each individual class are represented by the True Positive (TP) and True Negative (TN) values. These measures indicate if the preprocessing is favoring a single class or concept.

Table 3. Confusion matrix for performance evaluation of a binary classification problem

Actual	Predicted	
	Positive	Negative
Positive	True positive (TP)	False negative (FN)
Negative	False positive (FP)	True negative (TN)

Also, four metrics that describe both classes independently are obtained from it:

True Positive Rate, defined as $TPR = \dfrac{TP}{TP + FN}$, is the percentage of positive instances correctly classified.

True Negative Rate, defined as $TNR = \dfrac{TN}{FP + TN}$, is the percentage of negative instances correctly classified.

False Positive Rate, defined as $FPR = \dfrac{FP}{FP + TN}$, is the percentage of negative instances misclassified.

False Negative Rate, defined as $FNR = \dfrac{FN}{TP + FN}$, is the percentage of positive instances misclassified.

In order to evaluate the performance in imbalanced classification scenarios, more robust metrics which make use of these rates exist. Two of the most widely used are the Geometric Mean (GM) [23] and the area under the ROC curve (AUC) [24]. The former is defined in Eq. 1 and it attempts to maximize the accuracy of each one of the two classes at the same time. The latter is defined by the area under the curve given by the Eq. 2 and it evaluates which model is better on average, with a single measure.

$$GM = \sqrt{TPR * TNR} \tag{1}$$

$$AUC = \frac{1 + TPR - FPR}{2} \tag{2}$$

3.3   Infrastructure

Concerning the infrastructure used to perform the experiments, the Hadoop cluster at University of Granada was used. The cluster consists of fourteen nodes connected via a Gigabit Ethernet network. Each node has a Intel Core i7-4930K microprocessor at 3.40 GHz, 6 cores (12 threads) and 64 GB of main memory working under Linux CentOS 6.9. The infrastructure works with Hadoop 2.6.0 (Cloudera CDH5.8.0), where the head node is configured as NameNode and ResourceManager, and the rest are DataNodes and NodeManagers.

4   Experimental Results

In this section, the performances achieved by a Decision Tree classifier after applying independently each of the preprocessing techniques are presented.

The following tables show the average results of applying all the preprocessing methods (ROS, RUS, SMOTE-BD and SMOTE-MR)[1] on the three datasets for each oversampling percentage and number of partitions values (shown on Table 2). Where boldface indicates the highest value, and the best result for each dataset is underlined.

Tables 4, 5 and 6 present the obtained values considering the GM, TPR and TNR performance measures, respectively. In Table 4, no significant differences were found between the results for the same parameters configuration for each method. In general, it can be seen that all of techniques are scalable in quality regarding the partition numbers. This behavior was not expected for SMOTE-MR because the kNN of each positive instance is calculated over the data of its partition. Further investigation is ongoing to fully understand this result.

As mentioned in Sect. 3, we expected a performance improvement with the increase in the oversampling percentage. In our current datasets of study, no

---

[1] The SMOTE variants are abbreviated as "SMT-BD" or "SMT-MR" in all tables.

major variations in performance are seen, and if so, a small variation goes in the opposite direction, giving worse results (e.g. higgs dataset) for larger oversampling percentage.

Comparing with previous experience on small data [14], the different obtained results may be due to the data quality, which has not been assessed yet. The selected large datasets may have high redundancy and, therefore, the zones which need to be more strengthened, are being neglected or underestimated.

Table 4. GM average results for the four methods over the datasets for 32, 64 and 128 partitions and for 100, 150 and 200 oversampling percentage.

Part.	32				64				128			
Method	SMT-BD	SMT-MR	ROS	RUS	SMT-BD	SMT-MR	ROS	RUS	SMT-BD	SMT-MR	ROS	RUS
Perc. Dataset												
100 higgs	0.6462	0.6477	**0.6580**	0.6560	0.6468	0.6474	**0.6562**	0.6505	0.6479	0.6476	0.6561	**0.6568**
covtype7	0.9249	0.9242	**0.9363**	**0.9363**	0.9251	**0.9279**	0.9232	0.9271	**0.9306**	0.9274	0.9234	0.9258
susy	0.7650	0.7671	**0.7675**	0.7633	0.7671	0.7681	**0.7685**	0.7666	0.7643	**0.7671**	0.7670	0.7655
150 higgs	0.6172	0.6157	**0.6271**	0.6223	0.6213	0.6153	**0.6261**	0.6225	0.6182	0.6135	0.6253	**0.6266**
covtype7	**0.9250**	0.9232	0.9123	0.9123	**0.9341**	0.9249	0.9235	0.9221	0.9285	0.9273	0.9200	**0.9298**
susy	0.7528	0.7578	**0.7667**	0.7389	0.7595	0.7646	**0.7666**	0.7432	0.7607	0.7645	**0.7646**	0.7274
200 higgs	**0.6006**	0.5973	0.5334	0.5245	**0.6011**	0.5976	0.5337	0.5196	0.5907	**0.5984**	0.5339	0.5472
covtype7	**0.9236**	0.9190	0.9186	0.9110	0.9196	0.9195	0.9187	**0.9236**	**0.9229**	0.9206	0.9187	0.9080
susy	0.7350	0.7357	**0.7364**	0.7191	0.7335	0.7323	**0.7378**	0.7293	**0.7432**	0.7334	0.7391	0.7411

Table 5. TPR average results for the four methods over the datasets for 32, 64 and 128 partitions and for 100, 150 and 200 oversampling percentage.

Part.	32				64				128			
Method	SMT-BD	SMT-MR	ROS	RUS	SMT-BD	SMT-MR	ROS	RUS	SMT-BD	SMT-MR	ROS	RUS
Perc. Dataset												
100 higgs	0.6126	0.6181	0.7534	**0.7608**	0.6114	0.6150	0.7610	**0.7783**	0.6209	0.6158	**0.7601**	0.7517
covtype7	0.9396	0.9627	**0.9691**	0.9687	0.9622	0.9511	**0.9784**	0.9499	0.9416	0.9423	0.9773	**0.9812**
susy	0.8511	0.8241	0.8131	**0.8536**	**0.8282**	0.8220	0.8045	**0.8282**	0.8153	**0.8221**	0.7825	0.8086
150 higgs	0.8074	0.8071	**0.8323**	0.4822	0.7909	0.8067	**0.8336**	0.4777	0.8024	0.8172	**0.8348**	0.4929
covtype7	0.9802	0.9815	**0.9963**	0.8655	0.9835	0.9822	**0.9940**	0.8952	0.9750	0.9777	**0.9961**	0.9599
susy	0.7069	0.7225	0.7553	**0.9154**	0.7333	0.7555	0.7445	**0.9087**	0.7378	0.7598	0.7375	**0.9294**
200 higgs	0.8527	0.8559	**0.9236**	0.3003	0.8520	0.8532	**0.9234**	0.2940	0.8549	0.8538	**0.9231**	0.3347
covtype7	0.9922	0.9930	**0.9947**	0.8657	0.9901	0.9910	**0.9942**	0.8979	0.9844	0.9853	**0.9944**	0.8541
susy	0.6411	0.6443	0.6456	**0.9387**	0.6376	0.6375	0.6489	**0.9240**	0.6706	0.6467	0.6526	**0.9063**

Regarding runtimes, RUS and ROS have the best outperform due to the simplicity of their algorithms. Then follows understandably SMOTE-MR, due to the local nature of its approach. The partitioned data is processed locally in each Map, resulting in lower times. The SMOTE-BD presents the highest times of all the algorithms tested, using a distributed data approach, but giving an exact result. For instance, the runtime for SMOTE-MR versus SMOTE-BD for the higgs dataset, with 32 partitions, 150% of oversampling and k equals to 5, is four times faster. It is evident that a compromise between the runtimes and the model approximation has to be considered. At last but not least, another important factor to point out is related with the data intrinsic quality, that has not been evaluated here, but it is possible to be affecting our results with redundancy and noise.

Table 6. TNR average results for the four methods over the datasets for 32, 64 and 128 partitions and for 100, 150 and 200 oversampling percentage.

Part.	32				64				128			
Method	SMT-BD	SMT-MR	ROS	RUS	SMT-BD	SMT-MR	ROS	RUS	SMT-BD	SMT-MR	ROS	RUS
Perc. Dataset												
100 higgs	**0.6817**	0.6787	0.5747	0.5655	**0.6844**	0.6816	0.5658	0.5437	0.6764	**0.6811**	0.5663	0.5738
covtype7	**0.9120**	0.8874	0.9046	0.9051	0.8896	**0.9054**	0.8712	0.9048	**0.9202**	0.9135	0.8724	0.8734
susy	0.6878	0.7140	**0.7244**	0.6826	0.7105	0.7178	**0.7342**	0.7096	0.7173	0.7158	**0.7519**	0.7247
150 higgs	0.4721	0.4698	0.4725	**0.8031**	0.4882	0.4694	0.4703	**0.8111**	0.4763	0.4606	0.4685	**0.7965**
covtype7	0.8729	0.8684	0.8354	**0.9617**	0.8874	0.8709	0.8580	**0.9499**	0.8844	0.8795	0.8497	**0.9006**
susy	**0.8031**	0.7954	0.7783	0.5965	0.7873	0.7738	**0.7893**	0.6079	0.7847	0.7693	**0.7927**	0.5692
200 higgs	0.4230	0.4169	0.3081	**0.9161**	0.4242	0.4185	0.3085	**0.9183**	0.4085	0.4195	0.3087	**0.8946**
covtype7	0.8598	0.8505	0.8484	**0.9586**	0.8541	0.8532	0.8489	**0.9501**	0.8654	0.8603	0.8488	**0.9654**
susy	**0.8430**	0.8402	0.8398	0.5510	**0.8439**	0.8415	0.8388	0.5756	0.8237	0.8319	**0.8371**	0.6061

5 Conclusions and Future Works

In this work, a behavior analysis of the current preprocessing techniques for balancing Big Data was presented. Each solution use the MapReduce programming model through the Apache Spark framework, one of the most popular to deal with Big Data nowadays. Three of them were developed with a local approach and one with a global one. The main reason that motivated us was to evaluate if those methods perform as in the traditional data size problems.

Usually, in "small data" scenarios the performance of the SMOTE is better than the ROS and RUS proposals. Even though, our experiments results in Big Data do not show the same behavior. One cause could be the data quality. The current available big datasets may have presence of redundancy, noise, dispersion, among others factors which deteriorate the quality of the experimental results. Our intuition is that applying preprocessing techniques to balance the dataset is not enough, in which a previous data cleansing stage, may be necessary.

As future work, the development of hybrid models focused on the areas where resampling is specially needed will be carried out. In particular, the zones of interest are those with the presence of small disjuncts and overlapping. The proposal will prioritize the analysis of the local neighborhood of each minority class instance.

References

1. Chen, C.L.P., Zhang, C.-Y.: Data-intensive applications, challenges, techniques and technologies: a survey on big data. Inf. Sci. **275**, 314–347 (2014)
2. Prati, R.C., Batista, G.E.A.P.A., Silva, D.F.: Class imbalance revisited: a new experimental setup to assess the performance of treatment methods. Knowl. Inf. Syst. **45**(1), 247–270 (2015)

3. Dean, J., Ghemawat, S.: MapReduce: simplified data processing on large clusters. In: Proceedings of the 6th Conference on Symposium on Operating Systems Design and Implementation, OSDI 2004, vol. 6, p. 10. USENIX Association, Berkeley (2004)

4. Ramírez-Gallego, S., Fernández, A., García, S., Chen, M., Herrera, F.: Big data: tutorial and guidelines on information and process fusion for analytics algorithms with mapreduce. Inf. Fusion **42**, 51–61 (2018)

5. García-Gil, D., Luengo, J., García, S., Herrera, F.: Enabling smart data: noise filtering in big data classification. Inf. Sci. **479**, 135–152 (2019)

6. Chawla, N.V., Bowyer, K.W., Hall, L.O., Kegelmeyer, W.P.: SMOTE: synthetic minority over-sampling technique. J. Artif. Intell. Res. **16**, 321–357 (2002)

7. Fernandez, A., Garcia, S., Herrera, F., Chawla, N.V.: Smote for learning from imbalanced data: progress and challenges, marking the 15-year anniversary. J. Artif. Intell. Res. **61**, 863–905 (2018)

8. White, T.: Hadoop: The Definitive Guide, 4th edn. O'Reilly Media, Sebastopol (2015)

9. Zaharia, M., et al.: Resilient distributed datasets: a fault-tolerant abstraction for in-memory cluster computing. Presented as part of the 9th USENIX Symposium on Networked Systems Design and Implementation (NSDI 2012), pp. 15–28. USENIX, San Jose (2012)

10. Karau, H., Konwinski, A., Wendell, P., Zaharia, M.: Learning Spark: Lightning-Fast Big Data Analytics, 1st edn. O'Reilly Media, Sebastopol (2015)

11. Meng, X., et al.: MLlib: machine learning in apache spark. J. Mach. Learn. Res. **17**(34), 1–7 (2016)

12. Zaharia, M., et al.: Discretized streams: fault-tolerant streaming computation at scale. In: Proceedings of the Twenty-Fourth ACM Symposium on Operating Systems Principles, SOSP 2013, pp. 423–438. ACM, New York (2013)

13. Batista, G.E.A.P.A., Prati, R.C., Monard, M.C.: A study of the behavior of several methods for balancing machine learning training data. SIGKDD Explor. Newsl. **6**(1), 20–29 (2004)

14. López, V., Fernández, A., García, S., Palade, V., Herrera, F.: An insight into classification with imbalanced data: empirical results and current trends on using data intrinsic characteristics. Inf. Sci. **250**(20), 113–141 (2013)

15. Fernandez, A., del Rio, S., Chawla, N.V., Herrera, F.: An insight into imbalanced big data classification: Outcomes and challenges. Complex Intell. Syst. **3**(2), 105–120 (2017)

16. Basgall, M.J., Hasperué, W., Naiouf, M., Fernández, A., Herrera, F.: SMOTE-BD: an exact and scalable oversampling method for imbalanced classification in big data. J. Comput. Sci. Technol. **18**(03), e23 (2018)

17. SMOTE-BD Spark Package (2018). https://spark-packages.org/package/majobasgall/smote-bd

18. Maillo, J., Ramírez-Gallego, S., Triguero, I., Herrera, F.: kNN-IS: an iterative spark-based design of the k-nearest neighbors classifier for big data. Knowl.-Based Syst. **117**, 3–15 (2017)

19. SMOTE-MR source code (2018). https://github.com/majobasgall/smote-mr

20. Fernandez, A., Herrera, F., Cordon, O., Jose del Jesus, M., Marcelloni, F.: Evolutionary fuzzy systems for explainable artificial intelligence: why, when, what for, and where to? IEEE Comput. Intell. Mag. **14**(1), 69–81 (2019)

21. Lichman, M.: UCI machine learning repository (2013)

22. Gutierrez, P.D., Lastra, M., Benitez, J.M., Herrera, F.: SMOTE-GPU: big data preprocessing on commodity hardware for imbalanced classification. Prog. Artif. Intell. **6**(4), 347–354 (2017)
23. Barandela, R., Sánchez, J.S., García, V., Rangel, E.: Strategies for learning in class imbalance problems. Pattern Recognit. **36**(3), 849–851 (2003)
24. Huang, J., Ling, C.X.: Using AUC and accuracy in evaluating learning algorithms. IEEE Trans. Knowl. Data Eng. **17**(3), 299–310 (2005)

Data Analytics for the Cryptocurrencies Behavior

Eduardo Sánchez[✉], Jose A. Olivas, and Francisco P. Romero

Department of Information Technologies and Systems,
University of Castilla-La Mancha, Ciudad Real, Spain
Eduardo.Sanchez00@outlook.es, {JoseAngel.Olivas,
Franciscop.Romero}@uclm.es

Abstract. The cryptocurrencies are a new paradigm of transferring money between users. Their anonymous and non-centralized is a subject of debate around the globe that paired with the massive spikes and declines in value that are inherit to an unregistered asset. These facts make difficult for the common daily use of the cryptocurrencies as an exchange currency as instead they are being used as a new way to invest. What we propose in this article is a system for the better understanding of the cryptocurrencies economical behavior against the global market. For that we are using Data Analytics techniques to build a predictor that uses as inputs said external financial variable. These forecasts would help determine if a coin is safe to trade with, if those forecasts can be precise by only using this external data. The results obtained indicates us that there is a certain degree of influence of the global market to the cryptocurrencies, but that is it not enough to correctly predict the fluctuations in price of the coins and that they care more about others factors and that they have their own bubbles, like the crypto collapse in late 2017.

Keywords: Cryptocurrencies · Data analytics · Blockchain

1 Introduction

The cryptocurrency world is one of the most fascinating and unique paradigms that modern society is facing. It aims to change how money is generated and exchanged between people through what is called a blockchain. The main difference with classic currencies is the decentralization, most of the crypto coins are not associated to a single entity, corporation or country, although there are examples of the contrary. Said decentralization carries a degree of uncertainty, there is nothing behind the coin and the value of it depends entirely on the uses of the holders of the currency. It is also in a grey legal area, with more and more countries making steps for a more regulated crypto market.

So even though the original idea looks good in paper, there are many caveats left in the air that makes the trading with cryptocurrencies unsafe and unattractive for the general public because sometimes it feels that nobody has control on the value of them and overnight you could potentially lose a great part of your savings.

M. Naiouf et al. (Eds.): JCC&BD 2019, CCIS 1050, pp. 86–97, 2019.
https://doi.org/10.1007/978-3-030-27713-0_8

For this reason, we, the authors of this document, are going to develop a model to obtain knowledge and conclusions on the fluctuation on the price of the coins according to external economic circumstances. This way we can verify if the cryptocurrency behaviors go along the global economy or if they are independent, making them harder to predict than normal classic markets.

1.1 Data Analytics and Big Data

Data Analytics is the process of studying raw data sets for the extraction of knowledge [4, 6, 8]. There are many directions and techniques in which we can approach a data analytics project and one of them is what is called Big Data.

Big Data is none less than the treatment of large groups of data that are usually obtained in real time and need quick treatment so they can be useful. Economical models, such as this one, take advantage of the Big Data approach since using a predictor in real time adds a lot of value to the decision making that the end users have to affront. Making long-term determinations in investments is not as important as making short-term ones, and for that we need quick and fluent data processing.

1.2 Cryptocurrencies

One of the most interesting and controversial finance paradigms in modern society is the cryptocurrency boom. Between its novelty using current technology and exceptional situation and uses, the crypto phenomenon is a trending topic in both general public opinion and in the scientific community [2, 5, 10].

A cryptocurrency is a digital asset that uses cryptography and other encryption techniques to regulate the generation of those said assets and the verification of the transactions. (Mining) Said coins are stored, sent and received through a Cryptocurrency Wallet.

The initial idea for this new type of currency was so it was decentralised from any government, company or entity, for that it is the user computers that make the verification of the transactions.

This is done by the blockchain technology, developed by an anonymous person or group that went by the alias of Nakamoto [9] in 2009 for the developing of Bitcoin, the first cryptocurrency. This technology consists on the idea that everybody has a complete database of the blocks of the said blockchain. That way if everybody has the same data, said data must be true. This makes that all the transactions are public, but both the sender and the receptor are anonymous.

The workflow of the blockchain consists of that once a computer (node) receives a transaction it tries to solve a computationally difficult puzzle that once is done, if it is the first that solved it, it places the next block in the block chain and claims the rewards.

Those rewards are in fact a certain number of cryptocurrencies. The number of digital coins that provides as a reward depends on how many other nodes are mining that certain cryptocurrency and other factors that depends of the cryptocurrency.

There are a lot of different cryptocurrencies ready to be used. Alongside Bitcoin, the other most important and relevant cryptocurrency is Ethereum, but there is a plethora of smaller and less used coins that can be traded as well such as Monero, Ripple, Aion, Waves or Litecoin.

There have been some successes in the matter of correctly analysing and study the economical behaviour of the cryptocurrencies, such as an accurate forecasting on the price trend by the use of Gradient Boost Decision Trees [11] or a dissection of the bubbles in the Bitcoin history [12].

1.3   Workflow

This study explanation is divided by the following parts, starting in Sect. 2. In Sect. 2.1 we are going to define what is this project and what is not. For Sect. 2.2 we are explaining the caveats on the transformation of the cryptocurrencies timeseries to a characterization that allows us to work with it easily. Section 2.3 corresponds to the methodology for choosing which coins that represent the total the best are we going to study. Feature selection of the external economical variables used in the next stage for prediction is done in Sect. 2.4, and the prediction in Sect. 2.5. Evaluation of this prediction is detailed in Sect. 2.6.

Conclusions of the study are in Sect. 3 in which we talk about the discoveries of our investigation.

2   Analysis on the Cryptocurrencies Behaviour

The first step to take was the selection of what cryptocurrencies we were going to use to perform the study. This is a very important step since the goal is to obtain generalizations of the coins, not a case by case study. We knew that every coin was going to have different needs and different results, but if we could select a few ones that were representative of the conglomerate we can make conclusions of the general behavior of the cryptocurrencies. The selection was performed over 73 different coins which data was obtained from Coinmetrics.io, a webpage with the sole intent of providing raw data on cryptocurrencies and its related characteristics. Once we had the data, the selection was made as follows.

2.1   Scope of the Model

There are a lot of details to consider, an almost impossible task to cover in a single project, that is why in this document we are focusing on the relation between global economic variables and the cryptocurrencies price fluctuation. Using Artificial Intelligence, Data Analytics and Machine Learning the goal is to obtain knowledge about how related the fluctuations of the digital currencies with the global economy are, helping the interested user to forecast if it is safe to hold onto these new types of coins, and if so which of the coins are the safest. For that we are going to use supervised and unsupervised learning in conjunction. With unsupervised learning we are segmenting the coins to select the most relevant ones and generalizations. Supervised learning completes this process by taking those coins and using Machine Learning algorithms construct a predictor for each of them to forecast the value of them in the near future.

This model development is done by two Knowledge Discovery in Databases [7] processes that are done in order, not in parallel. The first one uses unsupervised learning to obtain this more relevant coins and the second one with supervised learning forecasts the value of the coins by only using external economic factors.

This knowledge is not aimed at investors, although it could be used by then, the goal is not to turn on a profit with this model, but to help final users make better decisions if they want to start exchanging money through this method.

2.2   Cryptocurrency Characterization

A cryptocurrency database is made by several characteristics, called features, such as the price or market cap, that had a value registered for each day from the moment the currency was coined to the last day that we stipulated to stop retrieving data in order to have the same end for every coin. This form to express data is called time series, and it is just a linear way to store information. It is widely used in economics where time and seasonality matters. For this project we used days as our grain, but we could had use months, semesters or even years, the timeseries model is not only thought to work with days [1]. Also, not all the coins had the same structure, some had more technical data like the block size, while others lacked it. We had to drop features and in some cases databases in order to have the same structure, so we could start working. Nonetheless, this structure was not useful to select the most representative coins, so we had to do a transformation and convert the data into one that can help us on the task.

Said transformation consisted into making a characterization of each coin considering the overtime price increases and decreases. The new database with all the characterization of the coins had the following features or variable

name: Name of the coin.

mean_daily_price_variation: Mean of the daily price variation in %.

cv_daily_price_variation: Coefficient of variation (relative standard deviation) of the daily price variation in %.

ratio_of_price_increases: From −1 to 1 it indicates the number of days that were positive of negative for coin, meaning that −1 indicates that all days were negative and 1 that every day was positive and 0 that there was an equal number of price increases and decreases.

linear_correlation_coefficient_price (Pearson's Correlation Coefficient): Measure of the strength of the linear relationship between number of days passed and price.

max_price_increase: Biggest increase for a day in %.

max_price_decrease: Biggest decrease for a day in %.

mean_active_addresses: Mean of the active addresses.

cv_active_addresses: Coefficient of variation (relative standard deviation) of the daily active addresses.

linear_correlation_coefficient_active_addresses (Pearson's Correlation Coefficient): Measure of the strength of the linear relationship between number of days passed and active addresses.

2.3   Representative Coins

Now that we had all the coins defined by one line in a database, we applied algorithms to make segments out of the entire database, and from there take the most representative coins. In this case we had chosen Principal Component Analysis, a statistical procedure to project a set of points into a reduced number of new linear uncorrelated values, to transform all these features into 2 data projections that we can show in a graph and K-means clustering [3] to obtain a determined number of clusters out of them. K-means clustering is a method of vector quantization that aims to divide several samples into a specified number, k, of clusters. Each sample belongs to the cluster with the nearest mean. With this algorithm we settled the number of clusters that we considered to be the best and obtained the most representative coins out of them. The k number of clusters at the start were 5 but got reduced to 4 since we got better results and the selected coins were Bitcoin, Ethereum, 0x, Icon and Waves (Fig. 1). The first two were selected since they are the most important cryptocurrencies at the time of writing this document, while the other three were the centroids of the resulting clusters. We did not select the 4th centroid of the last cluster since it was a very small cluster with just 2 coins and we considered that it was not going to contribute anything useful.

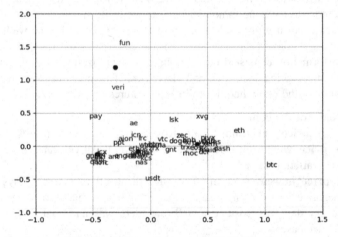

Fig. 1. KMeans clusters result with k = 4

2.4   Feature Selection

Once we had the coins selected, we had to choose the external economic factors from which we are going to base our model. After some investigation we concluded that the factors that were best suited for the problem were the price of Oil, Gold, the valuation of the market indexes Nikkei225, Financial Times Stock Exchange 100 and Dow Jones Industrial Average and the exchange in dollars of the Euro and Japanese Yen. The structure of the databases that contain this data is very similar to the cryptocurrencies databases, a timeseries with the grain put in days but instead of multiple features there was only one, the value in dollars.

What we did next was to finally calculate the importance that each of the selected coins gave to the external economic features. For that we used a correlation test to prune the economic features that were very similar considering the linear correlation with the market cap of the cryptocurrency. If two external variables acted identically, that could cause overfitting in out model, so we had to delete one of them. The one that was deleted was always whichever had the least linear correlation with the market cap of the coin at study.

Once we did this initial feature selection, we used a Random Forest predictor to calculate the importance. Random forests or random decision forests are an ensemble learning method for both regression and classification which build the model by using a N number of decisions trees for the training and making the model by selecting a portion of them. A decision tree is a tree-like flowchart used for decision support where each of the branches is a decision and the nodes are states. It can store a large amount of information such as chances that certain decisions are taken, cost of them and etcetera. Decision Tree learning implements that decision tree model to obtain conclusions about an item using the branches as observations. It is very useful specially to represent the decision-making process in data mining projects but can be also used as a predictor. Note that we are not using the Random Forest predictor to predict anything yet, we are only extracting the importance that it gives to each of the features while fitting the given data.

2.5 Behavior Prediction with External Data

To better validate that the obtained relevancies were on point, we developed a K-nearest neighbours model that considered the importance of each feature. For that we used a scaler to transform the numeric values of the global economic features into values from 0 to 1 and then multiplied then by the relevancy, which goes from 0 to 1 as well. After that transformation we operated like a normal predictor model, we parametrized the model by using a Cross-Validation test. Cross-validation (CV) is a statistical technique that help us determine the usefulness of a machine learning model. More in concrete, is a resampling procedure used to evaluate these models with preestablished data samples.

Once we parametrized the algorithm correctly, we divided the data into two sets, the training and test datasets. The training dataset is used to fit the model, to train it and adjust the predictor, and the testing dataset is used to validate and compare the predictions that the model can do with real data. Once the model is trained, we proceeded to forecast the price of the coins in 6 months for now, and that forecast was then compared with the testing dataset, obtaining the results showed in Figs. 2, 3, 4, 5 and 6.

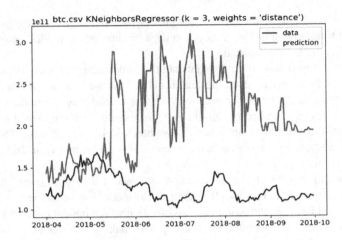

Fig. 2. Bitcoin six months prediction against real data

In the Bitcoin prediction we can observe that it gets accurate the first weeks, but as time passes, the normal behavior of the coin would be to go up in value. What the predictor cannot take into account are things like the collapse of the cryptocurrencies in the fall of 2017 and early 2018, which is an event completely unrelated with global economics.

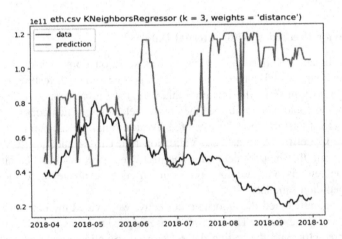

Fig. 3. Ethereum six months prediction against real data

Ethereum had the same type of result as Bitcoin, an overoptimistic prediction since the economical behavior of the coin did not follow the global market.

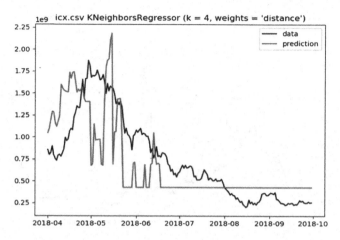

Fig. 4. Icon six months prediction against real data

Icon had an interesting prediction, it was not affected by the 2017 collapse like Bitcoin and Ethereum, this coin follows the global economy behavior much closer. The prediction falls down to a flat line due to the lack of samples that have that low of a value. This means the predictor could not tell that a new low was coming for this coin.

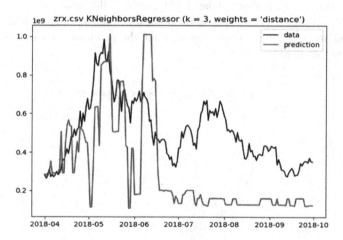

Fig. 5. 0x six months prediction against real data

0x does not follow the global economy as closer as the other coins, the prediction started to fail around summer, the slower months for economy. This did not hinder the movement of 0x.

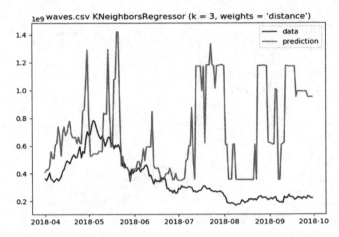

Fig. 6. Waves six months prediction against real data

Waves price does not follow the global market tendencies as we can see in the erroneous prediction.

2.6 Evaluation

The results obtained where then validated with another 3 different coins, Aion, Salt and Dash, each one from the 3 clusters resulting in the previous step. Said results were quite like their centroids counterparts so it proved that the predictor does not discriminate between the clusters of coins.

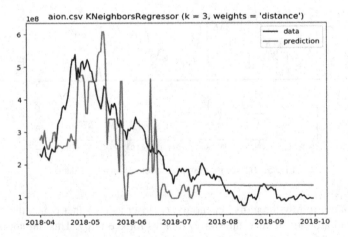

Fig. 7. Aion six months prediction against real data

Aion has similar behavior to Icon and 0x. The predictor can not estimate the new lows that a coin can have since it is out of the range of the training (Fig. 7).

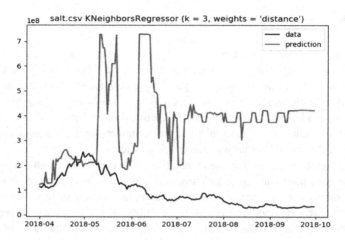

Fig. 8. Salt six months prediction against real data

As for Salt and Dash the predictor could not take into account the collapse of 2017, the same case as Bitcoin and Ethereum (Figs. 8 and 9).

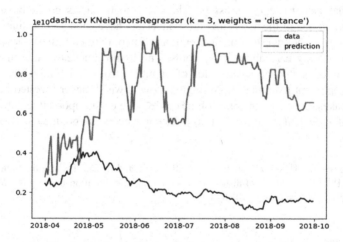

Fig. 9. Dash six months prediction against real data

Most of the predictions commence to err in the start of summer. This could mean that the predictor is not fit to forecast the price of a coin in a long-term situation of more than two or three months and that the economic slowness of these months do not affect the coins price.

3   Conclusions

What we learned is that both Bitcoin and Ethereum, the two largest and more known cryptocurrencies, follow the global market way closer than the rest of the samples. But at the same time, as we can see by looking at the predictions, these are also the two that are the hardest to predict with just these variables. This is not considered a contradiction, a coin can be very related to a certain feature, but without a global vision of the rest of the important features that affect it we cannot make a good estimation if the importance of the variables that we are studying is low.

What this really means is that even when those two cares a lot about the price of the oil, they are not conditioned by it and favor other features. Such features could be the number of miners currently mining the blockchain, the comments in social media and so on. It is as we explained earlier, from the 100% of the price, a Y percent is the economic external factors, from which that said percent, 80% is the value of the oil in dollars, but if the Y percent is really low, then it does not matter at all if the price of the crude has a lot of relevancy because in the grand scheme is not important at all.

The other three, smaller coins care a lot more about the economic features fluctuation since the prediction is more on point there, by a lot, and since they are varied in what features do, they give importance it indicates us that they are being traded by investors that also trade other assets. These cryptocurrencies could be used as a scapegoat when one of the assets that those investors have capital in declines in value, while the blob of the investors of Ethereum and Bitcoin are enthusiasts or specialized capitalists that only care about those coins. This is not unusual, since they are the better-known ones that attracted a lot of attention from tech individuals and just common people that liked the idea of digital money, which never invested in the past. That also makes sense why the price of the crude is the most important feature for those two, since it is the better indicator of how the home economy go, the more money that the common.

Acknowledgments. This work has been partially supported by FEDER and the State Research Agency (AEI) of the Spanish Ministry of Economy and Competition under grant MERINET: TIN2016-76843-C4-2-R (AEI/FEDER, UE).

References

1. Chatfield, C.: Time-Series Forecasting. Chapman & Hall/CRC, London (2001)
2. Wilson-Nunn, D., Zenil, H.: On the complexity and behaviour of cryptocurrencies compared to other markets. arXiv:1411.1924 [q-fin.ST] (2014)
3. Sutskever, I., Jozefowicz, R., Gregor, K., Rezende, D., Lillicrap, T., Orio Vinyals, O.: Towards principled unsupervised learning. arXiv:1511.06440, November 2015
4. Rouse, M.: SearchEnterpriseAI, November 2010. https://searchenterpriseai.techtarget.com/definition/AI-Artificial-Intelligence
5. Haferkorn, M., Quintana, J.M.: Seasonality and interconnectivity withing cryptocurrencies – an analysis on the basis of bitcoin, litecoin and namecoin. In: 7th International Workshop on Enterprise Applications and Services in the Finance Industry, FinanceCom 2014, Sydney, Australia, 22 January 2016

6. Elgendy, N., Elragal, A.: Big data analytics: a literature review paper. In: Perner, P. (ed.) ICDM 2014. LNCS (LNAI), vol. 8557, pp. 214–227. Springer, Cham (2014). https://doi.org/10.1007/978-3-319-08976-8_16

7. Fayyad, U., Piatetsky-Shapiro, G., Smyth, P.: The KDD process for extracting useful knowledge from volumes of data. Commun. ACM **39**(11), 27–34 (1996)

8. Perner, P. (ed.): ICDM 2014. LNCS (LNAI), vol. 8557. Springer, Cham (2014). https://doi.org/10.1007/978-3-319-08976-8

9. Nakamoto, S.: Bitcoin: a peer-to-peer electronic cash system, November 2008. https://bitcoin.org/bitcoin.pdf

10. Kim, Y.B., Kim, J.G., Kim, W., Im, J.H., Kim, T.H., Kang, S.J.: Predicting fluctuations in cryptocurrency transactions based on user comments and replies. PLoS ONE **11**(8), e0161197 (2016)

11. Sun, X., Liu, M., Sima, Z.: A novel cryptocurrency price trend forecasting model based on LightGBM. Finance Res. Lett. (2018, in press). Available online 27 December 2018

12. Aslanidis, N., Bariviera, A.F., Martinez-Ibañez, O.: An analysis of cryptocurrencies conditional cross correlations. Finance Res. Lett. **31**, 130–137 (2019)

Measuring (in)variances in Convolutional Networks

Facundo Quiroga[1]($\boxtimes$), Jordina Torrents-Barrena[2], Laura Lanzarini[1], and Domenec Puig[2]

[1] Instituto de Investigación en Informática LIDI, Facultad de Informática, Universidad Nacional de La Plata, La Plata, Argentina
`fquiroga@lidi.info.unlp.edu.ar`
[2] Intelligent Robotics and Computer Vision Group, Universitat Rovira i Virgili, Tarragona, Spain
`http://www.lidi.info.unlp.edu.ar/`
`http://deim.urv.cat/rivi`

Abstract. Convolutional neural networks (CNN) offer state-of-the-art performance in various computer vision tasks such as activity recognition, face detection, medical image analysis, among others. Many of those tasks need invariance to image transformations (*i.e.,* rotations, translations or scaling).

This work proposes a versatile, straightforward and interpretable measure to quantify the (in)variance of CNN activations with respect to transformations of the input. Intermediate output values of feature maps and fully connected layers are also analyzed with respect to different input transformations. The technique is applicable to any type of neural network and/or transformation. Our technique is validated on rotation transformations and compared with the relative (in)variance of several networks. More specifically, ResNet, AllConvolutional and VGG architectures were trained on CIFAR10 and MNIST databases with and without rotational data augmentation. Experiments reveal that rotation (in)variance of CNN outputs is class conditional. A distribution analysis also shows that lower layers are the most invariant, which seems to go against previous guidelines that recommend placing invariances near the network output and equivariances near the input.

Keywords: Transformation invariance · Rotation invariance · Neural networks · Variance measure · MNIST dataset · CIFAR10 dataset · Residual network · VGG network · AllConvolutional Network

1 Introduction

Convolutional neural networks (CNNs) provide outstanding results for several computer vision applications [4]. Nevertheless, CNNs have difficulty learning

M. Naiouf et al. (Eds.): JCC&BD 2019, CCIS 1050, pp. 98–109, 2019.
https://doi.org/10.1007/978-3-030-27713-0_9

good representations when objects appear rotated in many domains such as textures or galaxy classification, among other domains [4].

Dealing with rotations, or other transformations, requires the network f to be invariant or equivariant to the corresponding transformations Ts. Thus, f is invariant to T if altering the input x with T does not change the network output. In other words, $f(T(x)) = f(x) \forall x$. Alternatively, a network is equivariant to T if its output changes predictably when x is transformed by T. Formally, it is equivariant if there exists a function T' such that $\forall x$, we have $f(T(x)) = T'(f(x))$ [4]. Invariance is a special case of equivariance in which T' is simply the identity transformation. Analysing if f is equivariant to a T that operates on x requires finding a corresponding T' that operates on outputs [15]. Since CNNs are approximately invertible (sufficient condition) [7], the existence of T' is very likely. However, characterizing T' requires assuming its functional form and estimating its parameters [15].

Typical CNNs rely on feed-forward architectures with a series of convolutional layers followed by one or two dense layers. These models, commonly trained with stochastic gradient descent and without data augmentation, cannot learn invariances or equivariances to rotations [1,16]. Feed-forward networks exclusively composed of dense layers can approximate smooth functions given enough parameters. They can even learn arbitrary invariance and equivariance properties with heavy data augmentation [16]. Besides, some CNN models avoid dense layers [18] due to their questionable efficiency at dealing with certain inputs (e.g., images). On the contrary, convolutional layers, by definition, are translation equivariant and much more efficient, but they are not invariant nor equivariant to other transformations [4]. Since they have a lower representational power than dense layers, they cannot become so even with data augmentation.

Recently, models such as Transformation-Invariant Pooling [13], Deep Symmetry Networks [6], Steerable CNNs [3] were proposed to provide convolutional layers with rotation invariance or equivariance. Most schemes were based on modifying the filters so that they were invariant, or employing a set of equivariant filters which were subsequently pooled to supply invariance [3]. Other approaches made multiple predictions with rotated input versions to subsequently combine them [6]. For instance, Spatial Transformer Networks learned a canonical representation of the input [10].

Alternatively, data augmentation is also used to achieve partial invariance to geometric transformations of the input and improve generalization accuracy. While applied transformations are often mild, full rotation invariance is possible by including all transformation angles or deformations. This approach was studied for Deep Restricted Boltzmann Machines [14], HOGs and CNNs [15,16,19]. Although employed architectures are simpler, they generally require more training epochs to explore the wide space of rotated inputs. Thus, given sufficient computational budget, typical CNNs with data augmentation could learn the same set of filters that other models include by design [19].

In both cases, the network mechanisms to learn equivariant or invariant representations are not well-understood. It is still unclear whether the model or the data augmentation provides the invariance [3,10,13]. Many proposed layers are individually invariant or equivariant, but no analysis was reported to understand how networks as a whole encode such properties. Several authors studied which methods work best to achieve invariance [16,19], but no guiding principle in their analysis was employed except comparing the output accuracy. To the best of our knowledge, no works measure the internal invariance or equivariance of the network.

In this work, we shed some light on how and where CNNs represent and learn invariances. Notice that invariance can be readily estimated by measuring the changes of the network's outputs through the traditional *variance*. Following this principle, we define V to quantify the variance of a neural network with respect to a set of transformations. Our measure V can be computed not only in the output layers but in the internal layers and activations. This allows to visualize and quantify how invariant a network is as a whole and by layers or individual activations, thus providing insights about how invariance is encoded inside current CNNs. The method is applicable to any neural network, irrespective of its design or architecture, and any set of transformations.

2 Related Work

There are few works in the literature that measure invariance or equivariance in neural networks. Nevertheless, previous attempts to quantify invariance were mainly performed on translation and rotation transformations.

The work in [15] evaluated the equivariance of the internal convolutional representations with respect to a transformation T of the input. The proposed method assumed that the corresponding transformation T' acting on outputs was affine, and used empirical risk minimization to obtain the T' parameters once the network was trained. A different transformation T' was then estimated for each layer using the total network error as loss. A particular distance [15] from A_T to the identity matrix was utilized as an (in)variance measure of the layer's representation. Although this approach measures the equivariance, it *(i)* only deals with affine types, which limits its applicability to convolutional layers as a spatial correspondence for the affine map is needed, *(ii)* requires an arduous optimization process, and *(iii)* is not simple to interpret. Other works [2] modified the loss function to improve equivariance and invariance capabilities. However, the authors only estimated the loss impact with the technique of [15] in the last network layer.

Measuring invariance to transformations was also tackled from an adversarial perspective [5], confirming that simple rotations or translations can have a big impact on performance. In [11,16,19], the effect of using different data augmentation schemes and CNNs architectures was measured and compared. Specifically, the translation sensitivity map developed in [11] related the classifier accuracy with the center position of the object in the image. Equivalent 1D plots were employed in [11,16] to evaluate the rotation and other transformation invariances. Moreover, [1] studied the lack of equivariance in some CNNs by relating the Shannon sampling theorem to strided convolutions.

With the sole exception of [15], all aforementioned methods were focused on measuring how the network performance varies according to the learning algorithm, architecture or data augmentation scheme, disregarding the internal representation.

3   Proposed V Variance Measure

Invariance in neural networks is often measured just in the output layer by reporting the final performance (accuracy). Instead, we are interested in analyzing the invariance of the intermediate values or activations. Therefore, we propose a measure named V to quantify the (in)variance of the individual network activations with respect to a transform of the input. We assume that an activation is invariant when $V \sim 0$. By analyzing the activation's (in)variance, we can characterize the network's distribution of invariances in a fine-grained fashion. Variance at higher levels (*i.e.*, feature maps, spatial regions, layer types) can be calculated by aggregating the individual variances.

We denote the activation value as $a(x)$, where x is the network input. Note that a is not just an activation function such as *ReLU* or *tansig*, but an intermediate value or activation. Note that a may also be a resulting element of a matrix multiplication, convolutional filter, *ReLU* activation function, etc. To evaluate $V(a)$, we assume a finite set of samples X and transformations T. For example, X may be a set of images and T a set of rotations around the center of those images. The same definition works for other transformation and inputs.

We define the auxiliary set $a(X,T) = \{a(t(x)) \mid t \in T, x \in X\}$, which contains the activation values of a for all combinations of samples and transformations in X and T. Armed with this definition, we define the following (naive) way to calculate the variance (see Eq. 1):

$$V_{naive}(a, X, T) = Var(a(X,T)), \tag{1}$$

where $Var(X) = \frac{\sum_{x \in X}(x - Mean(X))^2}{n-1}$ and $Mean(X) = \frac{\sum_{x \in X} x}{n}$ are the standard sample variance and mean operators.

The main problem with this definition is that it mixes the variance generated by the randomness of the samples with the variance of the transformation. Instead, we calculate the variance generated by the transformation of a single sample, and then average the variance over the set of all samples (see Eq. 2):

$$V_{transformation}(a, X, T) = Mean(\{Var(a(x,T)) \mid x \in X\}) \tag{2}$$

While $V_{transformation}$ captures the variance we are interested in, comparing its value between different layers or even activations in the same layer is still hard, since values may differ significantly in scale. However, we can calculate a normalizing factor based on the variance computed over the samples (not transformations) (see Eq. 3):

$$V_{sample}(a, X, T) = Mean(\{Var(a(X,t)) \mid t \in T\}) \tag{3}$$

Equation 4 formulates the normalized transformation variance V:

$$V(a, X, T) = \frac{V_{transformation}(a, X, T)}{V_{sample}(a, X, T)} \tag{4}$$

Dividing by V_{sample} makes the magnitude order of V adimensional and comparable between layers. This expression is valid whenever $V_{sample}(a, X, T) > 0$ or $V_{sample}(a, X, T) = V_{transformation}(a, X, T) = 0$. Given a large enough X and T, in practice it is very hard to find cases for which both $V_{sample}(a, X, T) = 0$ and $V_{sample}(a, X, T) > 0$ hold. In those cases, however, we set $V(a, X, T) = +\infty$.

3.1 Stratified V

For categorization problems with a set of classes C, it is useful to measure the per class variance $V(a, X_c, T)$, where X_c is the set of samples belonging to class c. This shows if the invariance distribution is class specific.

We can also hypothesize that activation variances are different between classes, thus an alternative measure named $V_{stratified}$ can be defined (see Eq. 5):

$$V_{stratified}(a, X, T) = Mean(\{V(a, X_c, T) \mid c \in C\}) \tag{5}$$

Both $V_{stratified}(a, X, T)$ and $V(a, X_c, T)$ can assess if the invariance structure of a network is dependent on the class.

3.2 V for Convolutional Feature Maps

A convolutional layer outputs fm feature maps of size (w, h). The number of individual activations is $fm \times w \times h$, which can be too large. In those cases, we measure the variance of each feature map by aggregating the variance of the spatial dimensions. Given a feature map Fm such that $Fm(i, j)$ is the activation in the i, j spatial coordinates, we define $V_{fm}(Fm, X, T)$ as (see Eq. 6):

$$\begin{aligned} V_{f,transformation}(F, X, T) &= Sum(\{V_{transformation}(F(i, j), X_c, T) \mid \forall i, j\}) \\ V_{f,sample}(F, X, T) &= Sum(\{V_{sample}(F(i, j), X_c, T) \mid \forall i, j\}) \\ V_f(F, X, T) &= \frac{V_{f,transformation}(a, X, T)}{V_{f,sample}(a, X, T)} \end{aligned} \tag{6}$$

We aggregate the variances of the feature map with a Sum function, so that V represents the total variance. Aggregation is performed before normalization to remove the dependence on the size of the feature map with the division. Since feature maps are generally sparse, given that filters may be active only in certain spatial regions, aggregating activations with a $Mean$ function instead of Sum would significantly underestimate the variance of the feature map. Other aggregation functions such as Max suffer from similar problems.

3.3   Visualization of the Distribution of Invariances Across the Network via V

Calculating V on each activation independently enables the visualization of the distribution of invariances across the network at a glance. This performs qualitative analysis and guide research.

Activation variance can be visualized in various manners. For instance, Fig. 1 displays a heatmap of the activations' variance in a simple CNN trained on MNIST.

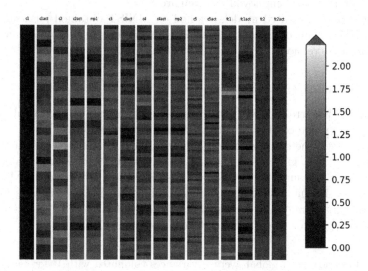

Fig. 1. Sample heatmap showing V for each activation of a CNN. Columns correspond to layers, and each column cell corresponds to a different layer activation. Greater values indicate more variance.

4   Experiments and Discussion

This section shows the experiments performed to validate the designed measure. The code to repeat the experimentation[1] and the library to calculate V[2] are publicly available online.

4.1   Methodology

The experiments were conducted on both CIFAR10 and MNIST datasets to provide an easy interpretation of the results. The following CNN architectures were selected to test the proposed approach: *(i)* a simple CNN as baseline *(ii)* VGG16 [17], *(iii)* AllConvolutional (without dense layers) [18], and *(iv)* ResNet

---

[1] https://github.com/facundoq/rotational_variance.

[2] https://github.com/facundoq/inmeasure.

(with residual connections and bottlenecks) [8]. We ignored the activations of the Batch Normalization layers [9] since these are linear transformations of the previous activations and therefore the variance is strongly correlated.

We assessed V on each model/dataset combination. For each combination, two networks were trained: i a rotated network with rotational data augmentation derived from random and uniform angles within the range 0°–360°, and ii an unrotated network with no data augmentation. Both were trained with the ADAM [12] optimizer using a learning rate of 10^{-3}. Rotated networks were trained for approximately twice the number of epochs than unrotated. Note that only test samples were employed to calculate V.

Rotated and unrotated models trained on MNIST obtained an accuracy of $\sim 99\%$. CIFAR10 unrotated models achieved accuracies between 65% and 75%, whereas rotated versions performed slightly worse (6% ↓).

4.2   Experiment 1: Validation of V

To validate if V correlates with the invariance of the model, we calculated the global average variance of both rotated and unrotated models.

Table 1 shows the obtained results for each pair. In all cases, the variance of unrotated models is greater, which confirms the viability of V as a transformational variance measure. For VGG and ResNet models trained on MNIST, the difference in V is small, which may be due to bigger models have a more complex representation and may capture finer detail. Indeed, both VGG and ResNet may be overly complex for the MNIST dataset.

Table 1. Comparison of global average variances computed with V for each pair of model and dataset. Greater values indicate less invariance.

Dataset	Version	AllConvolutional	SimpleConv	ResNet	VGG
MNIST	Rotated	0.47	0.59	0.70	0.68
MNIST	Unrotated	0.68	0.80	0.74	0.77
CIFAR10	Rotated	0.44	0.34	0.54	0.50
CIFAR10	Unrotated	0.73	0.65	0.82	0.66

Table 2 shows the obtained results with $V_{stratified}$. In general, variance values are significantly higher for the stratified version. This is because V_{sample} is lower, since by calculating the variance for each class independently, the inter-class variance of the representations is ignored. Therefore, networks learn different invariant representations for each class (*i.e.*, the invariance is class-specific).

4.3   Experiment 2: Variance Distribution Across Layers

Deep neural networks build increasingly higher level and more complex representations as the depth increases. It has been argued that there is a similar pattern

Table 2. Comparison of global average variances computed with $V_{stratified}$ for each pair of model - dataset. Greater values indicate less invariance.

Dataset	Version	AllConvolutional	SimpleConv	ResNet	VGG
MNIST	Rotated	0.77	0.92	0.97	0.97
MNIST	Unrotated	0.96	1.1	1.0	1.0
CIFAR10	Rotated	0.51	0.37	0.59	0.54
CIFAR10	Unrotated	0.83	0.71	0.89	0.72

with respect to equivariance and invariance for classification problems [4]; lower layers should be equivariant to preserve the multiplicity of representations, and the last layers should be invariant to perform the final classification.

We test this hyphotesis by calculating the average value of V for each of the layers of the models. We included all activations from all layers, except for BatchNormalization layers. Figure 2 shows such values for both stratified and normal variants of V, as well as rotated and unrotated datasets (blue and red, respectively).

For models trained with unrotated data, the variance increases quasi monotonically, which suggests that higher level activations are more susceptible to rotations of the input. Indeed, since these networks where not trained with rotated data, their resulting dynamics when tested on rotated data seems essentially random, with cumulative errors in representation building up through the layers.

For MNIST (Fig. 2(a) to (d)), the distribution of invariances is not significantly different between rotated and unrotated models, except for the last layers. These final layers must then code the invariance, as suggested by [4]. Nonetheless, this trend decreases as the model increases in complexity and depth (left to right). Indeed, ResNet models (Fig. 2(d)) show a significant decrease in variance beyond layer 35 for the rotated models.

For CIFAR10, however, models trained with rotated data show a general pattern of invariant lower layers, variant middle layers, and invariant final layers. This would suggest that equivariance in lower layers is not necessary or a pattern that emerges naturally. The middle layers are possibly coding equivariant representations of the objects, based on the more invariant features generated by lower layers.

In the case of the simple CNN (Fig. 2(a) and (e)), it clearly seen that the variance of an activation layer is always higher than that of the previous convolutional layer. This trend is still visible but less noticeably on the other models, in both datasets.

By comparing the values of the stratified and normal versions of V, the figures also confirm the relationship between both variants mentioned in Subsect. 4.2. Furthermore, we can see that $V_{stratified}$ is lower than V for all layers, not just globally.

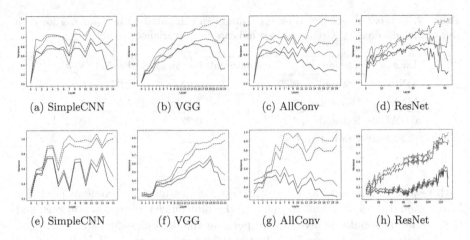

(a) SimpleCNN　　　(b) VGG　　　(c) AllConv　　　(d) ResNet

(e) SimpleCNN　　　(f) VGG　　　(g) AllConv　　　(h) ResNet

Fig. 2. Average V for different layers for MNIST (top row), CIFAR10 (bottom row), and all models. Dashed lines show results for models trained on unrotated datasets, while solid lines correspond to rotated datasets. Blue lines show results for the stratified version of V, while red lines indicate the normal version. (Color figure online)

4.4　Experiment 3: Class Conditional Invariance

We analysed $V(a, X, T)$ for the softmax output layer of the networks, but for each class separately. Figure 3 shows heatmaps of this variance for MNIST and Fig. 4 shows the same for CIFAR10. We used the normal, non-stratified variant of V, but the stratified version shows similar results.

In each heatmap, each column corresponds to the variance of the softmax, for the samples of different classes. That is, each column c shows $V(a_1, X_c, T)$, $\ldots, V(a_C, X_c, T)$, where X_c is the set of samples of class c, and $a_1, \ldots, a_C$ are the softmax outputs of the network. Therefore, the entry in row r and column c is the variance of the r element of the softmax when evaluated with samples of class c.

The first row of Fig. 3 shows that, for all models trained on rotated MNIST, the heatmap has a diagonal structure. This suggests that the softmax specific for a class is more invariant with respect to that class than all others. In other words, the invariance is class conditional; the networks are not learning to be invariant in all their outputs, just in those corresponding to the expected class. On the other hand, the output for class 1 results in a high variance for all classes, which indicates that the network has difficulties representing the digit 1 in rotated positions, not just recognizing it.

For unrotated MNIST (Fig. 3, row 2), this situation does not arise; the distribution of invariances shows that networks learn to be slighly invariant to class 0 because it has a natural rotational symmetry. The variance heatmaps for unrotated CIFAR10 (Fig. 4) show a similar lack of structure.

In the case of rotated CIFAR10, the heatmaps do not show a well defined diagonal structure. This could be due to the fact that models trained on

CIFAR10 only obtained about 70% accuracy, while on MNIST this was always approximately 99%. On the other hand, this phenomena could be due the networks being overfit on MNIST.

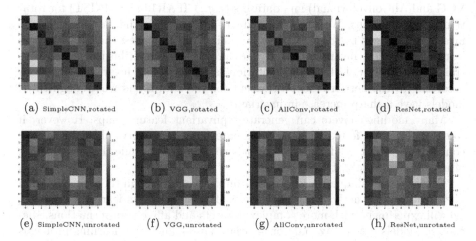

(a) SimpleCNN,rotated   (b) VGG,rotated   (c) AllConv,rotated   (d) ResNet,rotated

(e) SimpleCNN,unrotated   (f) VGG,unrotated   (g) AllConv,unrotated   (h) ResNet,unrotated

Fig. 3. Comparison of V for the output layer, by class, for MNIST (rotated/unrotated) and all models.

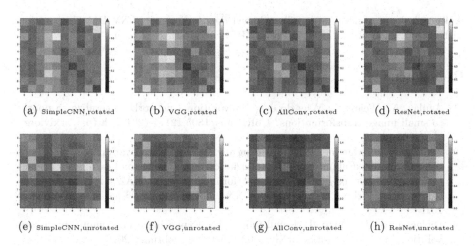

(a) SimpleCNN,rotated   (b) VGG,rotated   (c) AllConv,rotated   (d) ResNet,rotated

(e) SimpleCNN,unrotated   (f) VGG,unrotated   (g) AllConv,unrotated   (h) ResNet,unrotated

Fig. 4. Comparison of V for the output layer, by class, for CIFAR10 (rotated/unrotated) and all models.

5   Conclusion

In this work, we present a flexible, simple and understandable measure to quantify the (in)variance of individual CNN activations. Our measure is completely adaptable in terms of network architectures, layer types, arbitrary inputs and transformations. We also propose a modified version of the variance measure to analyze convolutional feature maps.

We evaluated the proposed measure V on well-known models (*e.g.,* ResNet, VGG and AllConvolutional) and datasets (*e.g.,* CIFAR10 and MNIST) for rotation transformations. We also validated the correlation of the measure with the invariances learned by the networks. Our main findings suggest that: *(i)* networks learn class specific invariances, and *(ii)* the invariance distribution of networks trained with/without data augmentation is similar. The invariance increases according to the network's depth, except for the final layers where augmented models have a sharp increase in invariance.

Many modified layers can generate equivariant feature maps. However, if some of them are found to be approximately invariant, it would suggest that equivariance is not needed. Thus, we believe that V will guide the development of new and more robust CNN models.

Future work will qualitatively assess our measure through the representation (*i.e.,* feature map activations and filters) of what networks learn. In addition, we will experiment with more complex datasets and affine transformations, specialized models with equivariant or invariant layers, and other learning problems (*e.g.,* segmentation). Finally, we will extend the experimentation to successfully describe the network invariance before training (random networks), and after training and fine-tuning.

Acknowledgments. We gratefully acknowledge the support of NVIDIA Corporation with the donation of the Titan X Pascal GPU used in this research.

References

1. Azulay, A., Weiss, Y.: Why do deep convolutional networks generalize so poorly to small image transformations? CoRR abs/1805.12177 (2018). http://arxiv.org/abs/1805.12177
2. Tensmeyer, C, Martinez, T.: Improving invariance and equivariance properties of convolutional neural networks. In: International Conference on Learning Representations (2017). https://openreview.net/forum?id=SyBPtQfAZ
3. Cohen, T.S., Welling, M.: Steerable CNNs. arXiv:1612.08498 [cs, stat], December 2016
4. Dieleman, S., De Fauw, J., Kavukcuoglu, K.: Exploiting cyclic symmetry in convolutional neural networks. arXiv:1602.02660 [cs], February 2016
5. Engstrom, L., Tsipras, D., Schmidt, L., Madry, A.: A rotation and a translation suffice: fooling CNNs with simple transformations. CoRR abs/1712.02779 (2017)
6. Gens, R., Domingos, P.M.: Deep symmetry networks. In: Ghahramani, Z., Welling, M., Cortes, C., Lawrence, N.D., Weinberger, K.Q. (eds.) Advances in Neural Information Processing Systems 27, pp. 2537–2545. Curran Associates, Inc. (2014). http://papers.nips.cc/paper/5424-deep-symmetry-networks.pdf
7. Gilbert, A.C., Zhang, Y., Lee, K., Zhang, Y., Lee, H.: Towards understanding the invertibility of convolutional neural networks. In: Proceeding of the Twenty-Sixth International Joint Conference on Artificial Intelligence (IJCAI) (2017)

8. He, K., Zhang, X., Ren, S., Sun, J.: Deep residual learning for image recognition. In: Proceedings of the IEEE Conference on Computer Vision and Pattern Recognition, pp. 770–778 (2016)

9. Ioffe, S., Szegedy, C.: Batch normalization: accelerating deep network training by reducing internal covariate shift. CoRR abs/1502.03167 (2015). http://arxiv.org/abs/1502.03167

10. Jaderberg, M., Simonyan, K., Zisserman, A., Kavukcuoglu, K.: Spatial transformer networks. arXiv:1506.02025 [cs], June 2015

11. Kauderer-Abrams, E.: Quantifying translation-invariance in convolutional neural networks. CoRR abs/1801.01450 (2018). http://arxiv.org/abs/1801.01450

12. Kingma, D.P., Ba, J.: Adam: a method for stochastic optimization. In: 3rd International Conference on Learning Representations, ICLR 2015. Conference Track Proceedings, San Diego, CA, USA, 7–9 May 2015 (2015). http://arxiv.org/abs/1412.6980

13. Laptev, D., Savinov, N., Buhmann, J.M., Pollefeys, M.: TI-POOLING: transformation-invariant pooling for feature learning in convolutional neural networks. CoRR abs/1604.06318 (2016). http://arxiv.org/abs/1604.06318

14. Larochelle, H., Erhan, D., Courville, A., Bergstra, J., Bengio, Y.: An empirical evaluation of deep architectures on problems with many factors of variation. In: Proceedings of the 24th International Conference on Machine Learning, ICML 2007, pp. 473–480. ACM, New York (2007). http://doi.acm.org/10.1145/1273496.1273556

15. Lenc, K., Vedaldi, A.: Understanding image representations by measuring their equivariance and equivalence. arXiv:1411.5908 [cs], November 2014

16. Quiroga, F., Ronchetti, F., Lanzarini, L., Bariviera, A.F.: Revisiting data augmentation for rotational invariance in convolutional neural networks. In: Ferrer-Comalat, J.C., Linares-Mustarós, S., Merigó, J.M., Kacprzyk, J. (eds.) MS-18 2018. AISC, vol. 894, pp. 127–141. Springer, Cham (2020). https://doi.org/10.1007/978-3-030-15413-4_10

17. Simonyan, K., Zisserman, A.: Very deep convolutional networks for large-scale image recognition. arXiv e-prints arXiv:1409.1556, September 2014

18. Springenberg, J.T., Dosovitskiy, A., Brox, T., Riedmiller, M.A.: Striving for simplicity: the all convolutional net. In: 3rd International Conference on Learning Representations, ICLR 2015. Workshop Track Proceedings, San Diego, CA, USA, 7–9 May 2015 (2015). http://arxiv.org/abs/1412.6806

19. Srivastava, M., Grill-Spector, K.: The effect of learning strategy versus inherent architecture properties on the ability of convolutional neural networks to develop transformation invariance. CoRR abs/1810.13128 (2018). http://arxiv.org/abs/1810.13128

Database NewSQL Performance Evaluation for Big Data in the Public Cloud

María Murazzo[1]([⊠]), Pablo Gómez[2], Nelson Rodríguez[1],
and Diego Medel[1]

[1] Departamento e Instituto de Informática - F.C.E.F. y N. - U.N.S.J.,
Complejo Islas Malvinas, Ignacio de la Roza y Meglioli,
C.P. 5402, Rivadavia, San Juan, Argentina
marite@unsj-cuim.edu.ar,
{nelson,dmedel}@iinfo.unsj.edu.ar
[2] Alumno Avanzado Licenciatura en Cs. de la Computación - F.C.E.F. y N. - U.
N.S.J., Complejo Islas Malvinas, Ignacio de la Roza y Meglioli,
C.P. 5402, Rivadavia, San Juan, Argentina
pablo.gomez.allende@gmail.com

Abstract. For very years, relational databases have been the leading model for data storage, retrieval and management. However, due to increasing needs for scalability and performance, alternative systems have emerged, namely NewSQL technology. NewSQL is a class of modern relational database management systems (RDBMS) that provide the same scalable performance of NoSQL systems for online transaction processing (OLTP) read-write workloads, while still maintaining the ACID guarantees of a traditional database system. In this research paper, the performance of a NewSQL database is evaluated, compared to a MySQL database, both running in the cloud, in order to measure the response time against different configurations of workloads.

Keywords: Big data · Cloud computing · Spanner · Google Cloud Platform

1 Introduction

At present databases are not only expected to be flexible enough to handle different variety of data formats, they are expected to deliver extreme performance as well as easily scale to handle big data. According to an estimate there are 2.5 quintillion bytes of data created each day, but that pace is accelerating with the growth of the Internet of Things (IoT) (Gubbi et al. 2013).

Last two years 90% of the data in the world was generated. This growth of storage capacity, leading to the emergence of data management systems where data is stored in a distributed way, but accessed as well as analyzed as if it resides on a single machine. Different types of data as well as continuity in data availability has become more important than ever and expects data to be available 24×7 and from everywhere.

Structured Query Language (SQL) (McFadyen and Kanabar 1991) became the standard of data processing because it contains such as data definition, data manipulation and data querying, all under one umbrella. RDBMS have always been distinguished

© Springer Nature Switzerland AG 2019
M. Naiouf et al. (Eds.): JCC&BD 2019, CCIS 1050, pp. 110–121, 2019.
https://doi.org/10.1007/978-3-030-27713-0_10

by the ACID (Atomicity, Consistency, Isolation, Durability) principle set that ensures that data integrity is preserved at all costs. RDBMS (Relational database management systems) can guarantee performance on the order of thousands of transactions per second, but in this time online transaction processing (OLTP) (Plattner 2009) in stages such as games, advertising, fraud detection and risk analysis involves more than million transactions per second that traditional RDBMS cannot easily handle. High availability without any point of failure as well as durability such as challenges have created a new wave processing database solutions, that manages data in structured unstructured ways.

New type of data management solutions are emerging to handle distributed content on open platforms. Unstructured data, non-relational databases, distributed architectures and big data analysis can change the way in which data is stored and analyzed, with the aim of obtaining useful information for making decisions in real time.

The objective of this paper is to address the problem of storage, recovery and analysis of large volumes of data using non-traditional databases.

The rest of this paper is organized as follows. Section 2 explains the generalities of big data, as well as the challenges that must be faced; Sect. 3 explains the Related technology and your convergence with big data; Sect. 4 addresses the generalities of the database administrators and in particular NewSQL solutions. In the following section the working environment and framework used is detailed. Finally, the developed tests, conclusions and future works are explained.

2 Big Data Characterization

Under the explosive increase of global data, the term of big data is mainly used to describe enormous datasets (Kacfah Emani et al. 2015). Big data is a term utilized refer to the increase in the volume of data that are difficult to store, process, and analyze through traditional database technologies. Compared with traditional datasets, big data often includes masses of unstructured data that need more real-time analysis. In addition, big data also brings new opportunities for discovering new values, helps us to gain an in depth understanding of the hidden values, and also incurs new challenges, ex. how to organize and manage such datasets (Oussous et al. 2018).

Nowadays, big data related to the service of Internet companies grow rapidly. For example, Google processes data of hundreds of Petabyte (PB), Facebook generates log data of over 10 PB per month, Baidu, a Chinese company, processes data of tens of PB, and Taobao, generates data of tens of Terabyte (TB) for online trading per day. While the amount of large datasets is drastically rising, it also brings many challenging problems demanding prompt solutions:

- The latest advances of information technology (IT) make it more easily to generate data. Therefore, we are confronted with the main challenge of collecting and integrating massive data from widely distributed data sources.
- The rapid growth of cloud computing and the Internet of Things (IoT) further promote the sharp growth of data. Cloud computing provides safeguarding, access sites and channels for data asset. In the paradigm of IoT, sensors all over the world are collecting and transmitting data to be stored and processed in the cloud. Such

data in both quantity and mutual relations will far surpass the capacities of the IT architectures and infrastructures of existing enterprises, and its real-time requirement will also greatly stress the available computing capacity. The increase growing data cause a problem of how to store and manage such huge heterogeneous datasets with moderate requirements on hardware and software infrastructure.

- In consideration of the heterogeneity, scalability, real-time, complexity, and privacy of big data, we shall the datasets at different levels during the analysis, modeling, visualization, and forecasting, so as to reveal its intrinsic property and improve the decisionmaking.

According to the analyzed, big data is an abstract concept. In general, big data shall mean the datasets that could not be perceived, acquired, managed, and processed by traditional IT and software/hardware tools within a tolerable time. Due to different concerns, scientific and technological enterprises, research scholars, data analysts, and technical practitioners have different definitions of big data. So, there have been considerable discussions between industry and academia about the definition of big data. In addition to developing a proper definition, the big data research should also focus on how to extract its value, how to use data, and how to transform "a bunch of data" into "big data" (Barrionuevo 2018). In that sense, NIST defines big data (NIST n.d.) as *"Big data shall mean the data of which the data volume, acquisition speed, or data representation limits the capacity of using traditional relational methods to conduct effective analysis or the data which may be effectively processed with important horizontal zoom technologies"*, which focuses on the technological aspect of big data. It indicates that efficient methods or technologies need to be developed and used to analyze and process big data.

2.1 Big Data Challenges

The sharply increasing of data in the big data's age brings huge challenges on data acquisition, storage, management and analysis. Traditional data management and analysis systems are based on the relational database management system (RDBMS). However, such RDBMSs only apply to structured data, others semi-structured or unstructured data. In addition, RDBMSs are using more and more expensive hardware. It is apparently that the traditional RDBMSs could not handle the huge volume and heterogeneity of big data.

In (Mukherjee et al. 2019) discuss obstacles in the development of big data applications. The key challenges are listed as follows:

- *Data representation:* many datasets have certain levels of heterogeneity in type, structure, semantics, organization, granularity, and accessibility. Data representation aims to make data more meaningful for computer analysis and user interpretation. However, an improper data representation will reduce the value of the original data and may even obstruct effective data analysis. Efficient data representation shall reflect data structure, class, and type, as well as integrated technologies, so as to enable efficient operations on different datasets.
- *Analytical mechanism:* the analytical system of big data shall process masses of heterogeneous data within a limited time. However, traditional RDBMSs are strictly

designed with a lack of scalability and expandability, which could not meet the performance requirements. Non-relational (NoSQL) databases have shown their unique advantages in the processing of unstructured data and started to become mainstream in big data analysis. Even so, there are still some problems of non-relational databases in their performance and particular applications. We shall find a compromised solution between RDBMSs and non-relational databases.

– *Expendability and scalability:* the analytical system of big data must support present and future datasets. The analytical algorithm must be able to process expanding and more complex datasets.
– *Cooperation:* analysis of big data is an interdisciplinary research, which requires experts in different fields cooperate harvest the potential of big data. A comprehensive big data network architecture must be established to help scientists and engineers in various fields access different kinds of data and fully utilize their expertise, so as to cooperate to complete the analytical objectives.

3   Related Technology

In order to gain a deep understanding of big data, is necessary introduce a fundamental technology that are closely related to big data: cloud computing. There is no a consensual definition of Cloud Computing yet. One of the most cited definition is the NIST's

(Mell and Grance 2011), where Cloud Computing is defined as being *"a model for enabling ubiquitous, convenient, on-demand network access to a shared pool of configurable computing resources (e.g., networks, servers, storage, applications, and services) that can be rapidly provisioned and released with minimal management effort or service provider interaction".*

Cloud computing is a technology to perform massive-scale and complex computing. It eliminates the need to maintain expensive computing hardware, dedicated space, and software. Big data is the object of the intensive computation operation and stresses the storage capacity of a cloud system. The main objective of cloud computing is to use huge computing and storage resources under concentrated management, so as to provide big data applications with fine grained computing capacity (Hashem et al. 2015). The development of cloud computing provides solutions for the storage and processing of big data. The distributed storage technology based on cloud computing can effectively manage big data; the parallel computing capacity by virtue of cloud computing can improve the efficiency of acquisition and analyzing big data.

Through the use of the cloud, access to software, hardware, and IaaS delivered over the Internet and remote data centers is possible. Cloud services have become a powerful architecture to perform complex large scale computing tasks and include a range of IT functions from storage and computation to database and application services. The need to store, process, and analyze large amounts of datasets has driven many organizations and people to adopt cloud computing (Liu 2013).

Cloud computing and big data are conjoined. Big data provides to users the ability to use commodity computing to process distributed queries across multiple datasets and return resultant sets in a timely manner. Cloud computing provides the underlying engine through the use of distributed data-processing platforms. Thanks to cloud, big data are stored in a distributed fault-tolerant database and processed through a programming model for large datasets with a parallel algorithm distributed in a cluster.

Cloud computing has a leverage effect on Big Data, providing the computing and storage resources necessary to Big Data applications. The inherent characteristics of Cloud Computing, such as elasticity, scalability, automation, fault tolerance, and ubiquity offer an ideal environment for the development of Big Data applications.

4 Big Data Database

One of the challenges that confront organizations dealing with Big Data is how and where to store the tremendous amount of data. In this context, the most widespread data management technology is relational database management systems (RDBMS). The data is stored in a structured way in form of tables or Relations. With advent of Big Data however, the structured approach falls short to serve the needs of Big Data systems which are primarily unstructured in nature. Increasing capacity of SQL although allows huge amount of data to be managed, it does not really count as a solution to Big Data needs, which expects fast reply and quick scalability (Pokorný 2015; Madden 2012).

To solve this problem a new kind of Database system called NoSQL (Han et al. 2011) was introduced to provide the scalability and unstructured platform for Big Data applications. NoSQL doesn't only stand SQL. NoSQL databases consist of a value pair key, documents, graph databases or wide column stores which do not have a standard schema which it needs to follow. It is also horizontally Scalable as opposed to vertical scaling in RDBMS. NoSQL provides great promises to be a perfect database system for Big Data applications; however that doesn't reach because of some major drawbacks like NoSQL does not guarantee ACID properties (Atomicity, Consistency, Isolation and Durability) of SQL systems. It is also not compatible with earlier versions of database.

This is where NewSQL (Kumar et al. 2014) is a latest development in the world of database systems. NewSQL is a Relational Database with the scalability properties of NoSQL. You can define NewSql as a next generation scalable relational database management systems (RDBMS) for Online Transaction Processing (OLTP) that provide scalable performance of NoSQL systems for reading writing workloads, as well as maintaining the ACID guarantees of a traditional database system.

4.1 Architecture NewSQL

Traditional databases cannot deliver capacity on demand that application development might be hindered by all the work required to make the database scale. To overcome scalability challenges, developers add scaling like separating, sharding and clustering techniques. Another common approach is to add larger machines at more cost. An ideal DBMS should scale elastically, allowing new machines to be introduced to a running database and become effective immediately. Therefore, To adopt scale-out performance,

DBMS that has been re-defined relational database technology and implement web-scale distributed database technology to tackle the multiple challenges associated with cloud computing and the rise of global application deployments (Moniruzzaman 2014).

Important Characteristics of NewSQL Solutions:

- NewSQL provides feature SQL as the primary mechanism for application interaction.
- NewSQL support ACID properties for transactions.
- NewSQL controls a non-locking concurrency control mechanism which is helpful for the real-time reads will not conflict with writes.
- NewSQL (dbShards) architecture providing much higher per-node performance than available fromtraditional RDBMS solutions.
- NewSQL support a scale-out, parallel, sharednothing architecture, capable of running on a large number of nodes without suffering bottlenecks.
- NewSQL systems are approximately 50 times faster than traditional OLTP RDBMS.

NewSQL databases provide an SQL query interface, and clients (users and applications) interact with them the same way they interact with relational databases. They manage read/write conflicts using non-lock concurrency control.

4.2   Selected Tool

There are many NewSQL databases being used in big data era, for real time web and big data applications. Every database is having a particular data format for storing its data. Hence a new customer faces the problem of selecting the appropriate NewSQL database that can be used for his business requirements, while migrating from relational database. On the other hand, there is another paradigm of shifting the big data applications from the physical infrastructure into the virtualized data centers in computational clouds. For this reason it has been decided to work with cloud computing as support architecture, more specifically Google Cloud Platform and Spanner as a tool.

Google Cloud Platform (GCP) (Google n.d.) is a suite of cloud computing services that runs on the same infrastructure that Google contained in Google's data centers. Alongside a set of management tools, it provides a series of modular cloud services including computing, data storage, data analytics and machine learning. In particular, Storage services which provides a variety of storage services, including: Cloud SQL (MySQL or PostgreSQL databases), Cloud Spanner and two options for NoSQL data storage (Cloud Datastore and Cloud Bigtable).

Cloud Spanner (Google 2017; Corbett et al. 2013) is a scalable, globally distributed database designed, built, and deployed at Google. At the highest level of abstraction, it is a database that shards data acros many sets of Paxos state machines in datacenters spread all over the world. Replication is used for global availability and geographic locality; clients automatically failover between replicas. Spanner automatically reshards data across machines as the amount of data or the number of servers changes, and it automatically migrates data across machines (even across datacenters) to balance load and in response to failures.

5 Proposed Work

In the present work, the performance comparison between a MySQL database and the Spanner database was performed, both working on the Google Cloud Platform.

Due to the need to work with large amounts of data, public data sets are used, in this case the data sets provided by the government of the Autonomous City of Buenos Aires are used. The first data set used contains data from the "Sistema Único de Atención Ciudadana" (SUACI), which is responsible for addressing the needs and claims of the residents of the CABA. This dataset is organized by year. For this work, only the year 2018 has been taken, which has 895,000 records with the following fields: contacto, periodo, categoria, subcategoria, concepto, tipo_prestacion, fecha_ingreso, hora_ingreso, domicilio_cgpc, domicilio_barrio, domicilio_calle, domicilio_altura, domicilio_es.

The second dataset used shows a register of the streets of the CABA, where the name, meaning and code of the streets are included among other data.

To perform the performance evaluation, three queries were defined:

- **Query 1:** `select count(*) from SUACI`
- **Query 2:** `select categoria, subcategoria, concepto, fecha_ingreso, fecha_cierre_contacto from SUACI`
- **Query 3:** `select RECLAMO.categoria, RECLAMO.concepto, RECLAMO.fecha_ingreso, RECLAMO.domicilio_calle as nombre_calle, CALLE.codigo as codigo_calle from SUACI RECLAMO, CALLEJERO CALLE WHERE RECLAMO.domicilio_calle = CALLE.nomoficial limit 11000;`

The three queries were executed with the same syntax in both databases because the implementation of the SQL language in both databases is similar, however it is necessary to emphasize that the data types used by MySQL are different from those uses Spanner, so the migration of data between both databases is not direct and may require adjustments.

The previously defined queries will be executed on three work scenarios: 2000, 10000 and 20000 records. This will allow to evaluate the behavior performance of MySQL and Spanner in front of the increase of workload.

To execute the queries to the MySQL database, the "MySQL Workbench" client was used, while to consult the Google Spanner database the console that provides the tool itself was used in the GCP.

In the case of the MySQL database, the way used to measure time is to connect the proxy "cloud_sql_proxy" provided by the GCP platform with the MySQL database through the console, then connect the MySQL Workbench client and subsequently execute the query and record the time (see Fig. 1). In the case of the Google Spanner database, the query is simply executed in the console provided by the tool and the time is recorded (see Fig. 2).

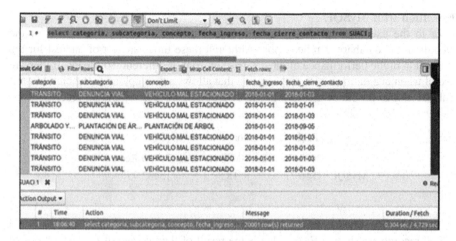

Fig. 1. Query execution in MySQL

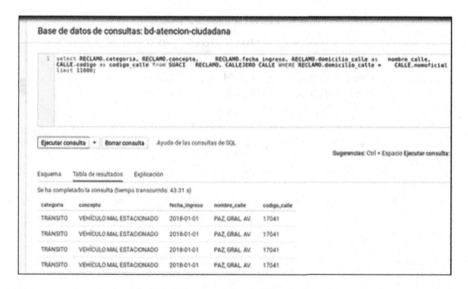

Fig. 2. Query execution in Google Spanner

For both cases, five measurements are made for each query and the average is calculated, which will be taken as the final result.

5.1 Results Analysis

The average times obtained for both databases are analyzed below. These times are expressed in seconds.

Execution with MySQL

Prior to the execution of the consultations it is necessary that the insertion of the records in the database is relied upon. Although these times are not of interest for the evaluation of the performance of the database in front of different workloads, if they are useful when evaluating which is the option that allows a data upload to the cloud faster. Table 1 shows the times incurred in the insertion of the corresponding records.

Table 1. Insertion times for MySQL

	2000 Records	10000 Records	20000 Records
MySQL	1856,77	9318,88	15482,95

In Table 2, the results of the executions of the three queries for 2000, 10000 and 20000 records are shown. Each value is the result of the average of five executions of each query. In Fig. 3, the results shown in Table 2 can be seen graphically.

Table 2. Execution times for MySQL

	2000 Records	10000 Records	20000 Records
Query 1	0,2494	0,2404	0,2666
Query 2	0,25	0,2624	0,3266
Query 3	0,2624	0,2678	0,2714

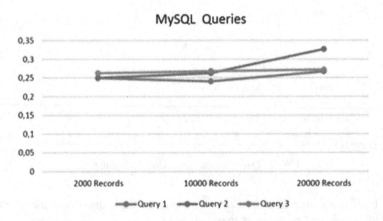

Fig. 3. Execution times for MySQL

Execution with Spanner

As in the case of the MySQL database, previous to the execution of the consultations it is necessary that the insertion of the records in the database is relied upon. Table 3 shows the times incurred in the insertion of the corresponding records.

Table 3. Insertion times for Spanner

	2000 Records	10000 Records	20000 Records
Spanner	34,21	39,72	41,75

In Table 4, the results of the executions of the three queries for 2000, 10000 and 20000 records are shown. Each value is the result of the average of five executions of each query. In Fig. 4, the results shown in Table 4 can be seen graphically.

Table 4. Execution times for Spanner

	2000 Records	10000 Records	20000 Records
Query 1	0,008878	0,03596	0,059554
Query 2	0,028214	0,054388	0,069064
Query 3	0,22888	0,24962	0,26448

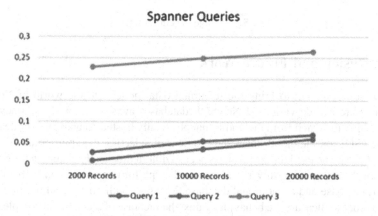

Fig. 4. Execution times for Spanner

Analyzing Tables 1 and 3 you can see that the insertion times of the registers in the database are significantly higher for MySQL. On the other hand, analyzing Tables 2 and 4, the best execution times for queries are obtained with Google Spanner. These results indicate that in the face of an increase in the number of registrations and the increased complexity of the queries, Google Spanner behaves more efficiently.

On the other hand, Table 5 summarizes the average times incurred by each manager in executing the queries. In this table it can be seen that in all cases Spanner has shorter execution times than MySQL, however for query 3 (the query with the greatest complexity because it contains a join) execution times are on average longer than for queries 1 and 2 (see Fig. 5). This leads us to the preliminary conclusion that Spanner has better performance than MySQL when the workloads are high, but this efficiency decreases when the complexity of the queries increases.

Table 5. Average execution times

	MySQL	Spanner
Query 1	0,2521	0,0348
Query 2	0,2797	0,0506
Query 3	0,2672	0,2477

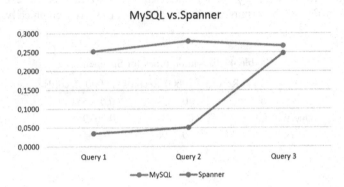

Fig. 5. Average execution times

6   Conclusion and Future Work

NewSQL solutions aim to bring the relational data model into the world of NoSQL. Since there are a wide variety of NewSQL databases available now days, a new customer wishing to switch from the traditional physically hosted database server design to using NewSQL on the cloud, faces the problem of selecting the appropriate NewSQL database that can be used for his business needs without drastically changing his existing application. That is why this research offers a performance evaluation between a NewSQL database and a traditional database, MySQL, both hosted and running in the cloud; in order to measure the behavior against the increase of records and complexity of the queries. According to the experiments performed, the use of Spanner offers better results than MySQL in the face of workload and complexity of the queries.

It is proposed as future work, the use of greater workload which will be implemented by increasing records in the database and increasing complexity in queries.

References

Barrionuevo, M., et al.: Estrategias y análisis orientados al manejo de datos masivos usando computación de alto desempeño (2018). http://sedici.unlp.edu.ar/handle/10915/68233

Marr, B.: How much data do we create every day? The mind-blowing stats everyone should read (n.d.).   https://www.forbes.com/sites/bernardmarr/2018/05/21/how-much-data-do-we-create-every-day-the-mind-blowing-stats-everyone-should-read/#61fd023f60ba. Accessed 18 Mar 2019

Corbett, J.C., et al.: Spanner: Google's globally distributed database. ACM Trans. Comput. Syst. (TOCS) **31**(3), 8 (2013). https://doi.org/10.1145/2491245

Google: Google Cloud Platform Overview—Overview—Google Cloud (n.d.). https://cloud. google.com/docs/overview/. Accessed 1 Apr 2019

Google: Cloud Spanner Documentation (2017)

Gubbi, J., Buyya, R., Marusic, S., Palaniswami, M.: Internet of Things (IoT): a vision, architectural elements, and future directions. Future Gener. Comput. Syst. **29**(7), 1645–1660 (2013). https://doi.org/10.1016/J.FUTURE.2013.01.010

Hashem, I.A.T., Yaqoob, I., Anuar, N.B., Mokhtar, S., Gani, A., Ullah Khan, S.: The rise of "big data" on cloud computing: review and open research issues. Inf. Syst. **47**, 98–115 (2015). https://doi.org/10.1016/J.IS.2014.07.006

Han, J., Haihong, E., Le, G., Du. J.: Survey on NoSQL database. In: 2011 6th International Conference on Pervasive Computing and Applications, pp. 363–366. IEEE (2011). https://doi.org/10.1109/ICPCA.2011.6106531

Kacfah Emani, C., Cullot, N., Nicolle, C.: Understandable big data: a survey. Comput. Sci. Rev. **17**, 70–81 (2015). https://doi.org/10.1016/j.cosrev.2015.05.002

Kumar, R., Gupta, N., Maharwal, H., Charu, S., Yadav, K.: Critical analysis of database management using NewSQL. Int. J. Comput. Sci. Mob. Comput. **35**(5), 434–438 (2014). http://s3.amazonaws.com/academia.edu.documents/33752078/V3I5201499a2.pdf? AWSAccessKeyId=AKIAIWOWYYGZ2Y53UL3A&Expires=1495314606&Signature= VhHyc%2BR0An%2FF8Oa6W5EAkCgxO9c%3D&response-content-disposition=inline% 3Bfilename%3DCritical_Analysis_of_Database_Management.pdf

Liu, H.: Big data drives cloud adoption in enterprise. IEEE Internet Comput. **17**(4), 68–71 (2013). https://doi.org/10.1109/MIC.2013.63

Madden, S.: From databases to big data. IEEE Internet Comput. **16**(3), 4–6 (2012). https://doi.org/10.1109/MIC.2012.50

McFadyen, R., Kanabar, V.: An Introduction to Structured Query Language. Wm. C. Brown, Dubuque (1991). https://dl.acm.org/citation.cfm?id=102896

Mell, P., Grance, T.: The NIST Definition of Cloud Computing (2011)

Moniruzzaman, A.B.M.: NewSQL: towards next-generation scalable RDBMS for online transaction processing (OLTP) for big data management (2014). http://arxiv.org/abs/1411. 7343

Mukherjee, S., Mishra, M.K., Mishra, B.S.P.: Promises and Challenges of Big Data in a Data-Driven World. In: Abraham, A., Dutta, P., Mandal, J., Bhattacharya, A., Dutta, S. (eds.) Emerging Technologies in Data Mining and Information Security. AISC, vol. 813, pp. 201–211. Springer, Singapore (2019). https://doi.org/10.1007/978-981-13-1498-8_18

NIST: NIST Big Data Working Group (NBD-WG) (n.d.). https://bigdatawg.nist.gov/home.php. Accessed 18 Mar 2019

Oussous, A., Benjelloun, F.-Z., Ait Lahcen, A., Belfkih, S.: Big data technologies: a survey. J. King Saud Univ.-Comput. Inf. Sci. **30**(4), 431–448 (2018). https://doi.org/10.1016/J. JKSUCI.2017.06.001

Plattner, H.: A common database approach for OLTP and OLAP using an in-memory column database. In: Proceedings of the 35th SIGMOD International Conference on Management of Data - SIGMOD 2009, pp. 1–2. ACM Press, New York (2009). https://doi.org/10.1145/ 1559845.1559846

Pokorný, J.: Database technologies in the world of big data. In: Proceedings of the 16th International Conference on Computer Systems and Technologies - CompSysTech 2015, pp. 1–12. ACM Press, New York (2015). https://doi.org/10.1145/2812428.2812429

Mobile Computing

A Study of Non-functional Requirements in Apps for Mobile Devices

Leonardo Corbalán$^{(\boxtimes)}$ (iD), Pablo Thomas (iD), Lisandro Delía (iD),
Germán Cáseres (iD), Juan Fernández Sosa (iD), Fernando Tesone (iD),
and Patricia Pesado (iD)

Computer Science Research Institute LIDI (III-LIDI),
School of Computer Science, National University of La Plata,
La Plata, Buenos Aires, Argentina
{corbalan,pthomas,ldelia,gcaseres,jfernandez,ftesone,
ppesado}@lidi.info.unlp.edu.ar

Abstract. Nowadays, no one questions the crucial role of Requirements Engineering in software systems development. Specifically, if apps are generated for execution on mobile devices, certain non-functional requirements become highly relevant. In this article, an experimental study on three non-functional requirements that are essential for the development of native and multi-platform mobile apps is detailed. These requirements are performance, energy consumption and storage space utilization.

Keywords: Native mobile apps · Multi-platform mobile apps ·
Non-functional requirements

1 Introduction

Not many years ago, Requirements Engineering was underestimated. Computer Science professionals considered design or coding as more challenging stages in software development. Currently, this situation has changed. Having the right requirements in early development stages considerably reduces the risk of problems appearing later on. In this context, and with the boom of mobile devices, software development is particularly conditioned by complying with certain requirements, some non-functional requirements in particular, that are critical in mobile apps.

In this article, the results obtained in [1–3] are expanded.

This study is aimed at quantifying the impact of the development approach used on three of the most popular non-functional requirements in the area of apps for mobile devices: performance, energy consumption and use of storage space.

To choose cases for study, the multi-platform development classification proposed by Raj and Tolety in [4] and reviewed by Xanthopoulos et al. in [5] was considered.

Computer Science Research Institute LIDI (III-LIDI)—Partner Center of the Scientific Research Agency of the Province of Buenos Aires (CICPBA).

M. Naiouf et al. (Eds.): JCC&BD 2019, CCIS 1050, pp. 125–136, 2019.
https://doi.org/10.1007/978-3-030-27713-0_11

This classification considers 4 categories: (1) mobile web approach, (2) hybrid approach, (3) interpreted approach and (4) cross-compilation approach.

The development approaches studied in this article were the native approach and the hybrid, interpreted, and cross-compilation multi-platform variations. The mobile web approach was excluded because a thorough analysis of that approach will be done in the future. The tests detailed in this article were run on the Android platform, which is the operating system that currently has the largest market share for mobile devices [6].

Even though the general design of the experiments revolves around the development approaches used, the tests had to be implemented using specific frameworks. There are several of these development frameworks for each of the approaches being considered. The ones chosen for this study are well known and very popular in the field. Based on these, six different analysis scenarios were defined: (1) Android SDK (native approach), (2) Cordova (hybrid approach), (3) Titanium (interpreted approach), (4) NativeScript (interpreted approach), (5) Xamarin (cross-compilation approach) and (6) Corona (cross-compilation approach). The same scenarios were used to test all 3 non-functional requirements being considered: performance, energy consumption and use of storage space.

It should be noted that the results presented in this paper are linked to the state of the art of the development framework used at the moment of carrying out the experiments and, therefore, could change in the future as these frameworks evolve.

The rest of this article is organized as follows: Sect. 2 discusses the issue of performance in mobile apps, Sect. 3 considers energy consumption, and Sect. 4 is devoted to the use of storage space. Finally, the conclusions and future lines of work are presented.

2 Performance

According to several quality standards, such as ISO/IEC 9126 and ISO/IEC 25010, an efficient performance (least possible processing time) is one of the attributes that any software application must meet [3–7]. This requirement gains significance in the context of mobile devices due to its direct relation to energy consumption [8], which is a critical aspect affecting battery autonomy.

User ratings in online app stores usually penalize low performance, generating negative publicity as a result [6]. Andre Charland and Brian Leroux identified processing time as one of the main issues to solve when developing multi-platform applications, and they stated that end users care about software quality and user experience [9].

Some authors have studied application performance based on the development approach used. The work done by Corral et al. [10] around native and hybrid apps (developed using Phonegap) for a version of the Android operating system can be mentioned. However, there are not many articles analyzing app performance based on the multi-platform development approaches mentioned in [4, 5] that were used as reference to establish the cases for analysis for this study.

2.1 Experiment

Tests were carried out for the 6 scenarios mentioned above, which include the most relevant development frameworks at the time of writing this article. For the tests, 3 different mobile devices were used – two smartphones and a tablet, all three with Android as OS; they are identified as Device A, Device B and Device C.

- **Device A**: *smartphone*, brand: Motorola, model: Moto-G2, processor: Quad-core 1.2 GHz Cortex-A7, RAM 1 GB Snapdragon 400, OS: Android 4.4.
- **Device B**: *smartphone*, brand: Samsung, model: S6, processor: Octa-core (4x2.1 GHz Cortex-A57 & 4x1.5 GHz Cortex-A53), RAM 3 GB Exynos 7420 Octa, OS: Android 5.0.2.
- **Device C**: *tablet*, brand: Samsung, model: Tab 2, processor: Dual-core 1.0 GHz, RAM 1 GB TI OMAP 4430, OS: Android 4.2.2.

A total of 18 test cases were defined – one per device per scenario.

To assess processing speed, the calculation of the following series was proposed:

$$serie = \sum_{j=1}^{5} \sum_{k=1}^{100000} \left(\log_2(k) + \frac{3k}{2j} + \sqrt{k} + k^{j-1} \right) \tag{1}$$

This expression includes several iterations, mathematical functions and f arithmetics. This type of calculation is frequent in applications that make intensive use of CPU computation power, such as games, augmented reality apps, image treatment apps, and so forth.

For each of the 18 test cases defined, 30 separate runs of the experiment were carried out, obtaining in each case a sample T, where $T = T_1, T_2, \ldots T_{30}$, and T_i = time required for calculating the series on the nth run of the experiment. Time T_i is expressed in milliseconds. In most of the practical cases of interest, 30 samples are enough to propose $\overline{T}$ as a good approximation to the real mean of the distribution. A large number of experimental works published in this field used this number of measurements for their data collection phases.

To characterize each of the samples obtained, statistic variables $\overline{T} = (1/n) \sum_{i=1}^{n} T_i$ and $S = \sqrt{\left(\frac{1}{n-1} \sum_{i=1}^{n} (T_i - \overline{T})^2 \right)}$, corresponding to the sample average and sample standard deviation, respectively, were calculated.

2.2 Results

Table 1 and Fig. 1 summarize the results obtained during the experiments. The values of $\overline{T}$ and S for the samples collected allow comparing the performance of the apps generated using the different development approaches in each of the devices used.

Clearly, for the three devices used for the tests, the apps generated with Native-Script, Titanium and Cordova were the most efficient ones, all of them completing the required calculation. On the opposite end, Xamarin, Android SDK and Corona always produced the slowest apps. Corona in particular stands out for its low performance.

Table 1. Processing time (ms) – Intensive calculation

Framework	Device A		Device B		Device C	
	$\overline{T}$	S	$\overline{T}$	S	$\overline{T}$	S
Android SDK	532.93	16.14	211.80	19.97	763.80	28.98
Cordova	230.33	14.22	85.77	8.83	190.60	9.36
Titanium	211.67	24.95	95.63	7.64	192.70	16.80
NativeScript	187.30	9.39	89.67	9.16	183.50	3.04
Xamarin	395.17	8.95	211.00	6.69	379.33	8.31
Corona	1401.73	12.60	600.53	5.95	1344.30	23.39

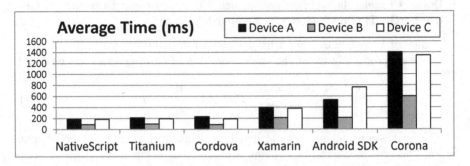

Fig. 1. Average processing time, in milliseconds.

It should be noted that the hybrid (Cordova) and interpreted (Titanium and NativeScrip) approaches, even if they operate differently, have one characteristic in common: they both run JavaScript code. In all these frameworks, the JavaScript V8 engine is responsible for optimizing the code that is then interpreted by a WebView. This engine has a crucial role and is largely responsible for the good results obtained. The tests for these approaches had a better behavior than that of the native approach (Android SDK) and the cross-compilation approach (Xamarin and Corona), which yielded the lowest performance.

The same relative differences were observed with the different development frameworks for all three devices used to carry out these tests.

3 Energy Consumption

Mobile device technology experienced a fast-paced development, significantly increasing its capabilities and performance, but also its energy requirements. Since battery technology has not evolved at the same speed, energy consumption has to be optimized to achieve a balance between performance and device autonomy.

Energy efficiency has become a relevant issue both for hardware manufacture as well as for software development. There is also a related requirement to protect the environment and general health of the planet. A higher energy consumption is against the current trend of *green computing*, which is attempting to achieve *eco-friendly* computer systems.

The introduction of ARM's big.LITTLE technology [11] to improve energy efficiency in mobile devices is an example of the commitment of hardware manufacturers with this issue [12]. However, the solution depends largely on software developers. It has been shown that, through changes in application source code, significant improvements can be obtained.

Many researchers have proposed solutions involving good programming practices. In [13], it was shown that the fastest algorithms are not always the ones that consume less energy. In [14] and [15], recommendations for the development of applications with a reduced energy demand were presented. In [16], it was concluded that most (5 out of 8) of the best programming practices published by Google to optimize Android app performance also had a positive impact on energy consumption. Other researchers explored the advantages of the technology called *Mobile Cloud Computing*, which integrates the concept of *cloud computing* to the mobile device environment. In [17] and [18], it was shown that this technology is effectively useful to save energy.

Energy efficiency and development *frameworks* for mobile devices have been widely studied separately; however, the articles considering both aspects simultaneously are scarce. This section is intended as a contribution in that direction, analyzing the effects of the development approach on app energy consumption. The 6 development frameworks mentioned in previous sections were considered, and 3 different types of common apps: (1) Intensive processing, (2) Video playback and (3) Audio playback.

3.1   Experiment

The platform chosen for testing was a medium-range smartphone – brand: Motorola, model: Moto-G2, processor: Quad-core 1.2 GHz Qualcomm Snapdragon 400, GPU: Adreno 305, RAM: 1 GB, OS: Android 6.0. This device was selected as an average representation of all devices considered during a preliminary testing phase.

Intensive processing, video playback and audio playback apps were developed, each of them in six different versions – one for each development *framework* being considered. This resulted in a total of 18 test cases. The intensive processing app consisted in calculating the series used for performance analysis, discussed above, represented by Eq. (1). The audio and video playback apps consisted in the playback of a one-minute long multimedia resource. In the case of the video, the file was 89.2 Mb in size, with a resolution of 1280×720 pixels, H.264 as codec at 5585 Kbps, and AAC audio tracks at 128 Kbps. For audio playback, the file used was 1.32 Mb in size and MP3 AC3 as codec, at 128 Kbps.

For energy consumption measurements, Qualcomm's *Trepn Profiler* tool was used; this is the same company that developed the smartphone processor used in all tests. To minimize external interference during the tests, a number of conditions were applied: (1) the device was on plane mode, (2) screen brightness was at the minimum level, (3) audio volume was set at 20%, (4) battery charge between 80% and 100%, (5) the device was not connected to the battery charger, (6) the app was implemented on dark mode, and (7) the app was running on the foreground, i.e., on the screen, during the test.

For each of the tests that were defined, 30 separate runs were executed, obtaining $X = X_1, X_2, \ldots X_{30}$ samples. In all cases, energy consumption, execution time and CPU percentage use were measured. Sample average $\overline{X} = (1/n) \sum_{i=1}^{n} X_i$ and sample standard deviation $S_X = \sqrt{\left(\frac{1}{n-1} \sum_{i=1}^{n} \left(X_i - \overline{X}\right)^2\right)}$ were calculated for all samples obtained.

3.2 Results

Tables 2, 3 and 4 summarize test results for intensive processing, video playback and audio playback, respectively. The histograms in Fig. 2 show sample distribution for energy consumption.

Table 2. Intensive processing app

Framework	Power (mWh)		CPU charge (%)		Duration (s)	
	$\overline{E}$	S_E	$\overline{C}$	S_C	$\overline{T}$	S_T
Cordova	1.597	0.136	35.924	2.571	8.467	0.679
Titanium	1.692	0.096	37.480	2.395	8.355	0.643
NativeScript	1.792	0.176	33.357	2.217	9.109	1.789
Xamarin	3.036	0.185	32.072	1.768	17.891	0.973
Android SDK	3.463	0.149	32.468	1.332	18.568	2.938
Corona	7.304	0.189	44.347	54.793	38.877	1.492

As regards intensive processing, there are three clear groups of *frameworks* (see Table 2 and Fig. 2 in Section A). The first group, with the highest energy efficiency, is formed by Cordova, Titanium and NativeScript. The second group, with medium efficiency, includes Xamarin and native Android SDK. Finally, with a much lower performance, Corona is in the group that has the greatest impact on battery autonomy. It is worth noting that the native development framework, Android SDK, is not among the most efficient approaches. This would be explained by the low performance of Java for mathematical functions in relation to execution time and, therefore, energy consumption.

As regards video playback apps, there are two clearly defined groups (see Table 3 and Fig. 2 in Section B). The first group, with greater energy efficiency, includes Android SDK, Corona, Xamarin and Titanium. The second group is formed by NativeScript and Cordova. These two frameworks showed a really low efficiency level, requiring more than double the power than the other frameworks that were tested. In particular, Cordova consumes almost triple the power than Android SDK (the best option). This significant difference is probably due to the fact that Cordova uses the HTML video player, which requires more CPU power than what the operating system provides.

In relation to audio playback apps, there are no major differences in energy consumption among the various development approaches with the exception of Corona,

Table 3. Video playback app

Framework	Power (mWh)		CPU charge (%)		Duration (s)	
	$\overline{E}$	S_E	$\overline{C}$	S_C	$\overline{T}$	S_T
Android SDK	4.776	0.287	14.540	0.862	61.600	0.814
Corona	4.992	0.235	14.704	0.711	62.733	0.907
Xamarin	5.119	0.473	15.465	1.608	62.333	0.959
Titanium	5.262	0.502	15.204	1.643	63.633	1.033
NativeScript	11.112	1.590	17.839	2.210	63.333	1.295
Cordova	13.866	0.536	22.358	0.903	62.833	0.834

Table 4. Audio playback app

Framework	Power (mWh)		CPU charge (%)		Duration (s)	
	$\overline{E}$	S_E	$\overline{C}$	S_C	$\overline{T}$	S_T
Android SDK	3.920	0.291	10.497	0.882	64.033	0.999
Xamarin	4.010	0.201	10.592	0.613	64.967	1.098
Titanium	4.189	0.277	11.865	0.835	64.767	1.104
NativeScript	4.224	0.229	11.233	0.644	65.867	1.042
Cordova	4.288	0.191	11.473	0.487	65.733	1.388
Corona	5.194	0.387	14.680	1.080	64.800	1.031

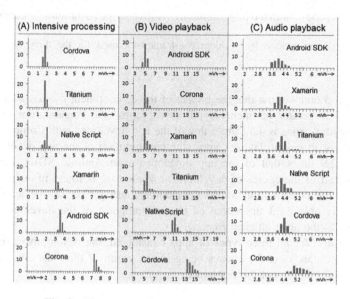

Fig. 2. Histograms representing test samples obtained.

which stands out as the least efficient of the bunch (see Table 4 and Fig. 2 in Section C).

Even though all audio and video tests were done using a resource that was exactly 60 s long, Tables 3 and 4 show differences in playback time. This is because the time required to start up the app is different for the different frameworks used. The native approach has advantages in this regard, but these become significant only if the app is used for a short time. As use time increases, the relevance of this advantage becomes less significant.

Figure 3 represents the values for $\overline{E}$ (sample average) obtained for the 18 test cases run. A quick visual inspection tells us that only one or tow of the development frameworks analyzed stand out for being inefficient (high energy consumption) for all three types of apps considered. This is the case of Corona for intensive processing, Cordova and NativeScript for video playback, and Corona for audio playback. On the other side of the coin, the most efficient framework does not stand out clearly from the other frameworks that are also efficient. This indicates that, even if there is no clear winner as the most efficient option, there is enough information as regards which frameworks should be avoided in each case.

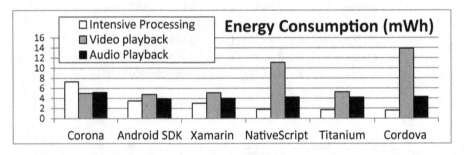

Fig. 3. Energy consumption by development framework used for the three types of apps considered.

Tables 2, 3 and 4 show that the impact of the development *framework* on energy consumption is greater in intensive processing apps (the consumption of the app developed with Corona is 4.57 times that of the app developed with Cordova). This is also significant in video playback apps (the consumption of the app developed with Cordova is 2.9 times that of the app developed with Android SDK). Finally, the smallest impact of the choice of development *framework* is found in audio playback apps, where the consumption of the development using Corona (the least efficient *framework*) is only 1.33 times that of the consumption of the development using Android SDK (the most efficient *framework*).

Additionally, for all three types of apps considered, the development framework with Titanium stands out for always being in the group of the most efficient frameworks, even if it never got the first place.

4 Storage Space

There is a lot of variation in storage space size among the different models of smartphones, this resource being critically scarce in less expensive devices. In the latter, the operating system and factory pre-installed apps take up a large portion of the available space, which hinders the installation of new apps [19]. This problem is worsened by a trend in the market towards the development of increasingly bigger apps.

According to a study carried out on Google Play Store, the size of the apps has quintupled between 2012 and 2017 [20]. This increase is due to the evolution of the market, requiring new features and better resources in apps. However, users are reluctant to resign storage space in their devices. The same study showed that the number of effective installations decreases by 1% for each 6 megabytes of increase in app size. Additionally, downloads are interrupted 30% more often in 100-megabyte apps than in 10-megabytes ones.

It is apparent that developers must optimize storage space usage to reach a larger number of potential users. Faced with this need, the scientific community has not been indifferent. In [21] and [22], elastic mobile app design models are proposed. These models use cloud computing technology to increase computation resources and storage space, splitting the apps into modules and migrating to the cloud those that require more resources. In addition to the obvious disadvantages, the excessive use of space can also negatively impact energy consumption [23]. In [24], methods to reduce the energy consumption associated with reading and writing access to storage space are proposed.

To minimize the size of the apps built, the impact of the development approach chosen on power consumption should also be considered. Below, we present the results of the experimental tests quantifying how large this impact is based on the development framework used.

4.1 Experiment

The tests whose results are presented in the following section were carried out with the 6 scenarios mentioned before. To assess the impact on the space used by the apps built, the size of the APK file generated by the different development frameworks being considered was measured. This information is independent from the device where the app is later installed.

The specific tools and libraries used by the development frameworks to support certain functionalities may impact differently the size of the resulting apps. To detect such potential situations, three different types of apps were implemented for each of the 6 scenarios defined, covering the usual functionalities: (1) text display, (2) video playback and (3) audio playback. Thus, there are 18 test cases: The source code for all developments produced for the experiments is publicly accessible on [25].

For all tests, the applications were generated following the standard procedure recommended by the documentation for each framework. In all cases, it was specifically corroborated that no additional files, such as images or videos, were included. These files are usually added by framework tools when a new app is built. Below, the results obtained are displayed.

4.2   Results

Table 5 and Fig. 4 show test results for the three types of apps considered. The development frameworks used are ordered based on the size of the APK file obtained. It can be seen that the sorting order is the same for all three types of apps considered. In all cases, the native development with Android SDK was the most efficient approach, producing the smallest app, followed closely by Cordova, hybrid approach. The apps generated with cross-compilation (Xamarin and Corona) are in intermediate positions. Finally, the frameworks using the interpreted approach (Titanium and NativeScript) generated the largest APK packages.

Table 5. Size (in Mb) of the app package produced

Framework	Text-based app	Video playback app	Audio playback app
Android SDK	1.48	1.04	1.38
Cordova	1.74	2.77	1.82
Xamarin	4.08	5.01	4.08
Corona	6.51	6.58	6.73
Titanium	8.54	9.23	9.16
NativeScript	12.49	12.47	21.11

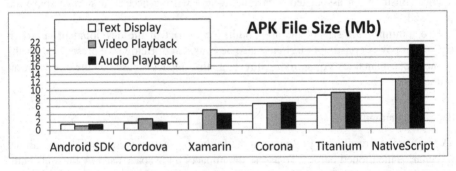

Fig. 4. Package size by development framework used for the three types of apps considered.

5   Conclusions and Future Work

In this article, the results obtained in our previous work, presented in [1–3], are expanded.

Thorough tests of three of the most important non-functional requirements were carried out. These requirements affect the development of apps for mobile devices: performance, energy consumption and use of storage space.

To analyze each of these requirements, 6 of the most popular development frameworks in the market were used: (1) Android SDK (native approach), (2) Cordova (hybrid approach), (3) Titanium (interpreted approach), (4) NativeScript (interpreted approach), (5) Xamarin (cross-compilation approach) and (6) Corona (cross-compilation approach).

The results obtained are a contribution for Software Engineers, allowing them to prioritize the use of an approach over others, based on the expected levels of performance, energy consumption and use of storage space.

Mobile app users heavily weight these non-functional requirements when it comes to deciding whether to install an app on their mobile devices. This work presents an advance in that regard, with concrete results.

Finally, an expansion is planned in the future to include iOS' mobile platform, in addition to carrying out tests with other frameworks for the native and multi-platform development approaches discussed here.

References

1. Delia, L., Galdamez, N., Corbalan, L., Pesado, P., Thomas, P.: Approaches to mobile application development: comparative performance analysis. In: 2017 IEEE SAI Computing Conference (SAI), pp. 652–659 (2017)
2. Corbalan, L., et al.: Development frameworks for mobile devices: a comparative study about energy consumption (ICSE). In: 2018 5th IEEE/ACM International Conference on Mobile Software Engineering and Systems on MobileSoft, Gothenburg, Sweden (2018)
3. Fernandez Sosa, J., et al.: Mobile application development approaches: a comparative analysis on the use of storage space. In: CACIC 2018, Tandil, Argentina (2018). ISBN 978-950-658-472-6
4. Raj, C.P.R., Tolety, S.B.: A study on approaches to build cross-platform mobile applications and criteria to select appropriate approach. In: 2012 Annual IEEE India Conference (INDICON), pp. 625-629. IEEE (2012)
5. Xanthopoulos, S., Xinogalos, S.: A comparative analysis of cross-platform development approaches for mobile applications. In: Proceedings of the 6th Balkan Conference in Informatics, pp. 213–220. ACM (2013)
6. Rösler, F., Nitze, A., Schmietendorf, A.: Towards a mobile application performance benchmark. In: ICIW 2014: The Ninth International Conference on Internet and Web Applications and Services, Paris, France (2014)
7. Jung, H.W., Kim, S.G., Chung, C.S.: Measuring software quality: a survey of ISO/IEC 9126. IEEE Softw. 21, 88–92 (2004)
8. Corral, L., Georgiev, A.B., Sillitti, A., Succi, G.: Can execution time describe accurately the energy consumption of mobile apps? An experiment in Android. In: GREENS 2014 Proceedings of the 3rd International Workshop on Green and Sustainable Software, pp. 31–37 (2014)
9. Charland, A., Leroux, B.: Mobile application development: web vs. native. Commun. ACM 54(5), 49–53 (2011)
10. Corral, L., Sillitti, A., Succi, G.: Mobile multiplatform development: an experiment for performance analysis. In: The 9th International Conference on Mobile Web Information Systems (MobiWIS), Ontario, Canada (2012)
11. big.LITTLE technology. https://www.arm.com/why-arm/technologies/big-little. Accessed Mar 2019

12. Banerjee, A., Roychoudhury, A.: Future of mobile software for smartphones and drones: energy and performance. In: Proceedings of the 4th International Conference on Mobile Software Engineering and Systems, pp. 1–12. IEEE Press (2017)
13. Bayer, H., Nebel, M.: Evaluating algorithms according to their energy consumption. Mathematical Theory and Computational Practice, p. 48 (2009)
14. Larsson, P.: Energy-efficient software guidelines. Technical report, Intel Software Solutions Group (2011)
15. Siebra, C., et al.: The software perspective for energy-efficient mobile applications development. In: Proceedings of the 10th International Conference on Advances in Mobile Computing and Multimedia, pp. 143–150. ACM (2012)
16. Cruz, L., Abreu, R.: Performance-based guidelines for energy efficient mobile applications. In: Proceedings of the 4th International Conference on Mobile Software Engineering and Systems, pp. 46–57. IEEE Press (2017)
17. Kumar, K., Lu, Y.-H.: Cloud computing for mobile users: can offloading computation save energy? Computer **43**(4), 51–56 (2010)
18. Gill, Q.K., Kaur, K.: A review on energy efficient computation offloading frameworks for mobile cloud computing (2016)
19. Vandenbroucke, K., Ferreira, D., Goncalves, J., Kostakos, V., Moor, K.D.: Mobile cloud storage: a contextual experience. In: Proceedings of the 16th International Conference on Human-Computer Interaction with Mobile Devices and Services (MobileHCI 2014), pp. 101–110 (2014)
20. Tolomei, S.: Shrinking APKs, growing installs, 20 November 2017. https://medium.com/googleplaydev/shrinking-apks-growing-installs-5d3fcba23ce2. Accessed Mar 2019
21. Zhang, X., Kunjithapatham, A., Jeong, S., Gibbs, S.: Towards an elastic application model for augmenting the computing capabilities of mobile devices with cloud computing. Mob. Netw. Appl. **16**(3), 270–284 (2011)
22. Christensen, J.H.: Using RESTful web-services and cloud computing to create next generation mobile applications. In: Proceedings of the 24th ACM SIGPLAN Conference Companion on Object Oriented Programming Systems Languages and Applications, New York (2009)
23. Lyu, Y., Gui, J., Wan, M., Halfond, W.G.J.: An empirical study of local database usage in android applications. In: IEEE International Conference on Software Maintenance and Evolution, Shanghai, China (2017)
24. David, G.Z., Nguyen, T., Qi, X., Peng, G., Zhao, J., Nguyen, T., Le, D.: Storage-aware smartphone energy savings. In: Proceedings of the 2013 ACM International Joint Conference on Pervasive and Ubiquitous Computing, New York (2013)
25. https://gitlab.com/iii-lidi/papers/apps-size.git. Accessed Mar 2019

Mobile and Wearable Computing
in Patagonian Wilderness

Samuel Almonacid[1,2]([✉]), María R. Klagges[3], Pablo Navarro[1,4],
Leonardo Morales[1,4], Bruno Pazos[1,4], Alexandra Contreras Puigbó[5],
and Diego Firmenich[1]

[1] Departamento de Informática (DIT), Facultad de Ingeniería,
Universidad Nacional de la Patagonia San Juan Bosco, Trelew, Chubut, Argentina
`informaticatw@ing.unp.edu.ar`
[2] Centro Para el Estudio de Sistemas Marinos (CESIMAR),
Centro Nacional Patagónico (CENPAT - CONICET),
Puerto Madryn, Provincia de Chubut, Argentina
`almonacid@cenpat-conicet.gob.ar`
[3] Instituto de Diversidad y Evolución Austral (IDEAus),
Centro Nacional Patagónico (CENPAT - CONICET),
Puerto Madryn, Provincia de Chubut, Argentina
[4] Instituto Patagónico de Ciencias Sociales y Humanas (IPCSH),
Centro Nacional Patagónico (CENPAT - CONICET),
Puerto Madryn, Provincia de Chubut, Argentina
[5] Departamento de Biología, Universidad Autónoma de Santo Domingo,
Santo Domingo, Dominican Republic
`http://www.dit.ing.unp.edu.ar`

Abstract. Recent advances in mobile and wearable technology in the
last few years have made the optimization of data collection processes
possible in diverse fields.

Users currently have access to small portable devices that are not only
sensitive to their activity, but also to their interaction with their environment.

These growing technological advances are in constant development,
and have given way to the study and redesign of processes that can
be tailored to fit any particular needs. Even users that are far from
urbanization, without access to electricity can make use of these possibilities. These technologies can substantially improve their productivity,
by allowing them to concentrate solely on their own tasks instead of on
the interactions with the computational method used to support their
activities. This study presents results and indicators relating to the application these tools within the field of Flora information retrieval, in areas
far from urban centers.

Keywords: Mobile computing · Wearable computing ·
Context-aware computing

M. Naiouf et al. (Eds.): JCC&BD 2019, CCIS 1050, pp. 137–154, 2019.
https://doi.org/10.1007/978-3-030-27713-0_12

1   Introduction

The smart mobile device user population has grown boundlessly in the last few years. Globally, not just in Argentina, the current availability and widespread access to these devices serves to forecasts the probability that in a few years (halfway through the next decade), every digitally active person will have these devices or likely even more powerful computational environments (see [1–3]).

Moreover, the way users interact with this technology evolves so fast that in most cases it is impossible to predict the environmental and social impacts it may cause.

The ubiquity levels reached by the scientific and the information industry in their applications have allowed a massive expansion in terms of what a common user can virtually do with their own devices. A new demand for software products that are ready-made for using these capabilities can be seen as a possible result of this evolution.

On the other hand, given that the methods, techniques, and technologies involved in the development of these kinds of products have been in constant growth for the last few years, there is a wide range of possibilities relative to the development of platforms. Being completely up to date in every project can sometimes be a challenging task for professionals and students within this area of specialization.

In this paper, we present not only the results of applying this technology to a highly versatile domain, but also the benefits obtained from implementing this technology to the flora surveying methods tasks done by botanologists in our community.

Results were obtained after four years of research, with designs and development in mobile and wearable software [4]. The impacts of applying this technology to the subject matter was then measured through a series of controlled experiments performed by an interdisciplinary biologist and programmer team.

Due to the encouraging results obtained from the experiments, these technologies were then incorporated by the biologists for their use in real case scenarios, and this allowed us to compare real results with the experimental results obtained from the previous phases.

The next chapter describes the domain of the aforementioned application, in order to facilitate an in depth comprehension of further sections. Chapter three describes the application developed, while chapter four summarizes the experimental results obtained with these applications. Finally, chapter five compares the experimental results with the values obtained from their use in real case uses. Conclusions and future works can be found in the last section of this document.

2   Application Domain

In terms of what the application domain is, this one in particular is based on vegetation surveys. While many different methods of surveying, measurement and ecological analysis exist, the need to rely on practical information is very important in order to obtain results that match up with reality as precisely as possible [5]. An adaptation of the "Point Quadrant" method (detailed in [6]) was used to adequately study the patagonian pasturelands, because this is a non destructive objective technique that easily allows for the evaluation of the different plant communities. This method is based on registering along a horizontal transect line, divided in 100 equidistant points, every single time plant species come into contact with a needle's tip as it is being let down in each point [6,7].

It should be noted that this method was implemented in remote areas far removed from urbanization, where the possibility of accessing an internet connection or even electricity is null. Biologists carry out three to six day surveying campaigns. During these campaigns they travel to the different points of interest within the area in one or more vehicles and perform an intensive supervised data collection on the area. In each transect at least 100 elements are registered concerning the species of flora found. The distance between the points of each transect depends on the estimated vegetation coverage. In areas where the coverage is greater than 65%, it is recomended to record each step, between 45% y 65% every two steps and less than 45% every three steps, the latter being the most common [7]. The first point would be located at 200 m from the roads and fencing in order to lessen the border effect produced by fragmentation [8]. Besides the location of the transect, information pertaining the state of soil, the species, notes and pictures are also registered. Each person in charge of carrying out the survey, in deviating 200 m from the established pathways, must also walk at least the complete length of the 300 m transect and return to the vehicle; in doing so, they would have traversed approximately 800 m in total for every survey (500 m from the first point to the last point of the transect and an additional 300 m to walk back to camp, the latter only if a shorter route back was available).

During campaigns the survey work starts at dawn and ends at sundown in order to maximize the amount of work that can be done under natural lighting, which is why the number of transects that can be successfully completed in a single day varies depending on the season it takes place in. It goes without saying that the person in this scenario has undergone taxing physical activity. In order to paint a clearer picture concerning survey campaigns, in just three days a single biologist will have walked 25 km, away from commodities like running water or comfortable sleeping conditions, and having to work in low temperatures between 3 and 12 °C.

The purpose of illustrating the work conditions for these particular campaigns is to give a graphic example on how these common occurrences in fieldwork translate over to real life examples and how this can lead any scientist to commit mistakes while working under strain. These mistakes can range from forgetting the exact mnemotechnic code for a particular transect, or even missing the first point by 15 km in unfamiliar terrain because they activated the wrong transect in the application to begin with.

2.1   Development

Persistence is a basic need for Data Collection. Initially, the data was collected using handwritten notes in templates printed for that purpose. Each sheet contains the data collected for one transect, which are then transcribed digitally to a spreadsheet.

This method possesses a series of drawbacks for the scientist, given that each printed sheet consists of a grid where columns represent points in the transect (one hundred columns) and rows represent species found during the trip. While there is no fixed value of species, the number of species found is typically close to twenty five, thus approximately thirty rows are printed in the template (see Fig. 1)

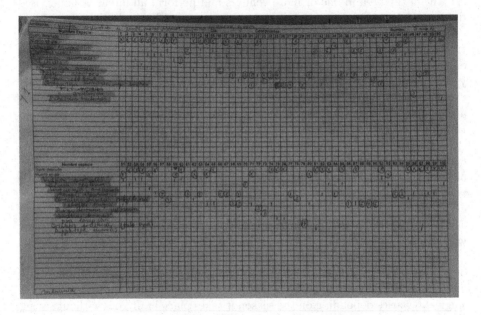

Fig. 1. Handwritten template

The probability for human error increases as the regular long hours of field-work progress, and as such the opportunity for committing mistakes like misplacing data in surrounding rows or columns also grows.

The same situation can occur while transcribing the sheets, where the scientist is forced to maintain high levels of concentration in order to avoid making any possible mistakes, which can produce a significant amount of physical and mental stress.

Moreover, as each cell has a very reduced area available, if more space is required to record more information for any given data collection point, it would have to be recorded elsewhere (like a second sheet), with a specific reference to the data collection point it belongs to.

In order to provide support in this task, the applications have been developed to make use of a computing area known as *Context-Awareness*, which in turn enables the change to go not just from a paper form to a spreadsheet, but rather a fully functional tool that can assist the data entry process completely.

A *Context-Aware System* is defined in [9] as a system that *[...] adapts according to the location of use, the collection of nearby people, hosts, and accessible devices, as well as to changes to such things over time.*

In the same way, *Awareness* and *Context* are defined by [10] as *the use of context to provide task-relevant information and/or services to a user* and define context as *any information that can be used to characterize the situation of an entity, where an entity can be a person, place, or physical or computational object* respectively.

Based on these definitions, different categorizations of Awareness can be applied. Table 1 describes the type of awareness implemented for the applications developed in this study.

Table 1. Implemented awareness in each application.

Sensibility	Sensing method	Application
Server Detection	Wi-Fi	LeafLab
Mobile Application Detection	Wi-Fi	LeafLab Wear
Adaptive Behavior	Continual survey of the species found	LeafLab and LeafLab Wear
Smart Suggest	Analysis of the same transect in previous visits	LeafLab Wear
Screen related Information (Help)	Based on the user in app section location	LeafLab
Location Sensitivity	GPS Sensor	LeafLab and LeafLab Wear
Orientation Sensitivity	Compass or Geomagnetic Sensor	LeafLab
Hand Gesture's Sensitivity	Accelerometer	LeafLab Wear
Energy Level Sensitivity	Internal System Sensor	LeafLab and LeafLab Wear

In order to arrive at a viable solution, an application was developed (LeafLab [11]) centered around data gathering and persistence phases, and was divided in to two modules: LeafLab Mobile and LeafLab Server.

Based on the results yielded by the first version of this application, a second version (LeafLab Wear, see [12]) was elaborated with more context-awareness, as well as the possibility of adding a new device (a Smart-watch), which greatly improved the efficiency of the system, and the user's experience. Both of these versions are briefly detailed below.

2.2   Leaflab

LeafLab Mobile is a hybrid mobile application (see [13–15]) developed to run in both Android and iOS based devices. The application was designed to be used in tablets, with the purpose of bringing the user all the relevant information without the need to navigate through numerous amounts of screens.

Context-Awareness was the cornerstone of this project, which meant making the application responsive to a specific set of conditions, such as the entirety of the transect, if the location corresponded to the data to be collected, device's details such as battery consumption and so on.

The application is composed by four general views (see Fig. 2). The first view depicts all the general data about the actual transect (amount of data already collected, remaining transect length from navigation point, direction, and fast access to typical tasks that can be done during this phase (to add an point of interest or note, take a picture of the place and attach it to the transect, etc.).

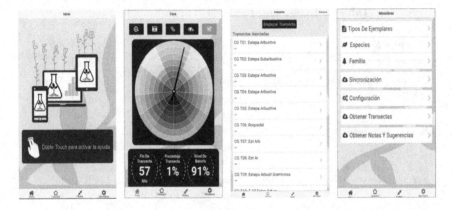

Fig. 2. Principal views of LeafLab. From left to right: Initial view (no transects activated), General information and navigation of a transect, List of collection campaigns and Miscellaneous.

In the next three views, the user can review the data that was collected in previous campaigns (collection of transects), the geolocation points in the actual transect, and to configure the set of relevant elements to be added in each point/transect (in addition to flora information).

The first step (if the application already has the information about the species to be collected), is to create and/or activate a transect. The shortcuts for the previously mentioned tasks will be made available from the principal view solely from this moment on.

To create a transect, the user can start it (create the first point) wherever they consider best, and the application will record that exact position to assist the user in future trips.

If the user activate a previously created one (which means, is a new visit over an already existent transect), the application will guide the user to the initial point, and will automatically activate the Information Load view when the user reach it.

The application will automatically maintain an up-to-date count of the completed points, so that the probability of inputting erroneous information between adjacent points is greatly reduced.

If the user needs to add more information it can be done through the same action menu by attaching it to the point or transect (depending on what user requires), without the need for any additional tools (except for the collection of physical samples), thus allowing a reduction in the time needed to accomplish these kinds of tasks when compared to the traditional methodology that uses physical template sheets.

In the other hand, *LeafLab Server* is a standalone application (the principal view of a transect can be seen in Fig. 3), developed to synchronize with the mobile pair and allows the user to have all the data that was collected in the campaigns to be backed up into a database for that purpose.

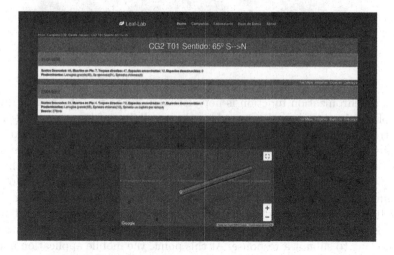

Fig. 3. Principal view of LeafLab Server for a transect.

In addition to the data persistence, *LeafLab Server* can visualize the information stored in the database distributed in a map according to their corresponding real position (see Fig. 4), and access detailed information related to it, which allows the user to query for species related items, and also gives them the possibility of downloading all the information in the template-based data sheets format that they are more familiar with.

The map makes use of the Google Map API (Application Programming Interface), and was implemented with the sole purpose of providing support for the user in the selection of the transect or campaign being revised, through an easy to read spatial reference that could recognizes when regions had been previously visited or not. While application's map function does make use of the Google Map API, it does not provide any type of integration with GIS (Geographic Information System) tools at this stage.

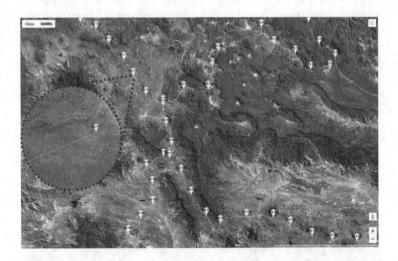

Fig. 4. View of initial points of some transects on the map.

The synchronization function is not just intended to free the user of several hours of digitization, but also to reduce the possibility of human error (caused by distraction, tiredness or stress) during the task. The synchronization was implemented using the two-phase commit protocol, in which a coordinator (in this case, the mobile app) initiates the synchronization process, while the rest of the participants (the LeafLab Server) are responsible for the transaction and informing the coordinator whether or not the request was successfully completed (by adding the data sent by the mobile application). The coordinator will abort the transaction if the server sends an Error Message warning, or if the servers take too long to submit a response. At this point, the mobile application notifies the user if the operation succeeded or failed, where they can then choose whether to make another attempt immediately or complete the synchronization at a later time.

While it is possible to use and synchronize the data base with data collected between multiple devices, this version of the software does not support concurrent connections, which means that the user can currently only synchronize one device at a time.

2.3   LeafLab Wear

In the same way LeafLab Mobile was focused in minimizing the manual effort needed for the data to be collected, LeafLab Wear was an Android Native project (developed making use of specific technology for Android based devices, see [13,14]) centered in the user experience [16,17], drastically improving the transparency and efficiency of the system.

This project was based on a research about users cases, interaction, and time consumption in the previous version of LeafLab.

With this in mind, the LeafLab Wear implementation expands the previously described system with the addition of a Smart Watch application, making use of their capabilities to minimize the time the user must spend interacting with the tools that are meant to be used as support.

LeafLab Wear relies on fast and simple user gestures [18] to diminish direct interaction with the mobile application, and therefore allowing the scientist to focus in their job instead of on their tools.

Furthermore, new features like communication between devices (Smart Watch and Tablet), smart suggestions based on the analysis of previously collected data from the same point, and a fast selection function that is triggered via the stimuli detected with the Smart Watch were also added.

The more frequently used functions by the scientists (such as selecting the specific species found in a point, see Fig. 5) can be quickly accessed via a list shown in the hand-held Smart Watch or even with a simple gesture, so the tablet just need to be used for more complicated or hardware specific tasks (like take pictures).

Shortening the user interaction time window with the tablet device provided the convenience of a longer battery life, which was especially beneficial given that accessing electrical energy is something of an impossibility at the locations where the surveys take place. This improves the overall work environment, because it frees the user from the constant concern that the device will run out of energy before the task can be fully completed.

In order for the application to display the desired characteristics, it became necessary to implement other important aspects such as Distributed User Interfaces and coordination aspects between both devices by modifying the first version of the application, which had originally only been contemplated to work independently on its own (see [19,20]).

The interfaces were semantically distributed (as detailed in [19,21]), which means that the elements the user can interact with in both the Tablet and the SmartWatch are represented in such a way that their meanings remain the same across platforms, even if it looks different from one device to another. Some of the wearable device's views can be seen in Fig. 6.

The next section details the results obtained from different experiments, as well as a comparison between the traditional methodology, and the first version of LeafLab and LeafLab Wear.

3   Preliminary Experiments

The experiments detailed in this section correspond to tests made to both LeafLab versions before their use in production.

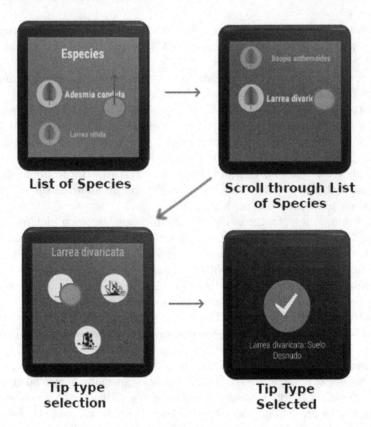

Fig. 5. Steps taken to complete a simple survey point.

3.1   LeafLab

In order to measure the performance of the mobile application in a real case, a two phase experiment was designed taking the following

- Localization: The tests were done in regions close to Telsen-Chubut (Provincial Rute 4), which presents the normal conditions for the use of the application.
- Time Span: The experimentation process lasted three days, requiring two days for the first phase, and one day for the second phase. The phases were executed in different periods of the year.
- Amount of data collected: The tests were executed using data corresponding to ten transects. Due to the aim of each phase, one thousand points were collected and analyzed for the first phase, whilst a series of measures were collected over each of the ten transect's initial points were collected for the second one.

Fig. 6. Some of the views implemented for the wearable device.

- Test team composition: The team was composed by three developers, and a biologist from the Universidad Nacional de la Patagonia "San Juan Bosco"'s Biology department, who was only familiarized with the traditional methodology, and had no background knowledge whatsoever regarding the application.

This way, the user had to learn how the application worked during the survey of the transects, and the team could collect data about the system's learning curve.

In general terms, the measures taken correspond to the total time needed to complete a transect, average time to complete a point, fastest survey method (the traditional way or with the application), and the time required to reach the first point of a transect.

3.2   LeafLab Wear

In order to verify the performance of the second version of the application with the wearable component in a real time scenario, an experiment was designed taking into account the following aspects:

- Localization: With the objective generating comparable results, the tests were performed in the same location and using the same transects as in the experimentation for the first version.

– Time Span: The time needed for the whole experimentation phase was one day (from 5 AM to 9 PM), on February of 2017.
– Amount of data collected: The same transects used during the experimentation for the first version were revisited (10 transects, 1000 points).
– Test Team Composition: Again, the team was composed by three developers and a biologist, but this time the biologist was already familiarized with the first version of the application, but with no previous knowledge of the wearable application.

As in the previous case, the user had to learn how the system worked with the new device (Smart Watch) during the survey, which sped up the work flow process because of how intuitive and easy to use the device application was.

In addition to the aforementioned measurements, the number of times the mobile application was used to register a point, the number of times the smart watch was used for the same task, the number of times the smart watch could have been used, and remaining energy in both devices before and after the trip was recorded.

4   Experimental Results

This section depicts the results of tests performed for each version of the application. Table 2 shows the time that takes the user to reach the first point of a transect, starting from the moment the applications were ready to guide the user. At first glance, version one of the application seemed to have a better performance, however the time lost in geolocation data acquisition and delays produced by choosing a wrong transect (the user must manually choose and activate each transect) was not measured. Taking these kind of problems into account, the average time for the first version ascended to 4′42″ (four minutes and forty two seconds, complete data in the Real Case Scenario section). The second version did take this into account, making the device sensitive to the actual location in a continuous way, so that the user doesn't need to wait for a GPS acquisition delay (when the vehicle stops moving, the device already has the user's geolocation). This way, the possibility of activating the wrong transect is also eliminated because the device automatically chooses the transect that is in closest proximity. This implies that in the second version, the total time necessary starts from the moment the vehicle stops until the user arrives at the first point, in contrast with the first version where the user was more likely to forfeit several minutes while the vehicle was no longer in motion or due to any of the previously explained situations.

Table 3 displays a comparison between the traditional methodology and the first version that was developed, depicting the points that were finished in the least amount of time by each method.

For the second version only the time required to complete the task using the application was recorded. Given that it was established that the use of the application itself did not delay the task, the improvement gained by using it was measured instead. There was approximately a 35% increase in the speed of the data collection.

Table 2. Time it takes to reach first point of transect

Transect	Time to first point	
	LeafLab	LeafLab + Wear
1	2' 30"	6' 35"
2	0' 53"	1' 50"
3	1' 00"	1' 46"
4	1' 12"	2' 25"
5	0' 55"	3' 46"
6	0' 40"	4' 26"
7	0' 34"	3' 34"
8	1' 23"	1' 50"
9	0' 51"	1' 32"
10	0' 38"	1' 43"
Total	10' 36"	29' 27"

Table 3. Comparison between LeafLab v1 and the traditional methodology

Transect	Less time with	
	Traditional	LeafLab (V1)
1	31	69
2	35	65
3	36	64
4	32	68
5	39	61
6	33	67
7	33	67
8	36	64
9	38	62
10	26	74
Total	33.9%	66.1%

5 Real Case Scenario

In the summers of 2016 and 2017, the previously mentioned tools were put to use in an actual campaigns, providing the opportunity to collect a great deal of data relevant to the established parameters (Table 4).

Figure 4 is a map of the geographic distribution of the 200 transects that were visited during the years 2016 and 2017, using both of the developed versions of Leaflab. The first version was used during the 2016 campaigns. During the experiment, sensitivity to context along the transect path was very limited

Table 4. Total time to complete a transect

Transect	Total time	
	LeafLab + Traditional	LeafLab (V2) + Wear
1	97′	39′ 25″
2	61′	32′ 48″
3	50′	32′ 04″
4	38′	31′ 00″
5	44′	28′ 17″
6	49′	33′ 49″
7	38′	25′ 12″
8	30′	25′ 46″
9	44′	38′ 20″
10	40′	29′ 36″
Total	491′	216′ 17″
Point AVG	45″–47″	26″–30″

because it had no previous records of the plant species found there due to it being the very first trip. In the course of the following year's campaigns, the second version of the application (in conjunction with LeafLab Wear), used the 2016 context data related to the plant species found in each transect of the campaigns in order to heighten the sensitivity to context information while the user walked along each transect line. Figure 7 depicts the progress of the tools used during the campaigns.

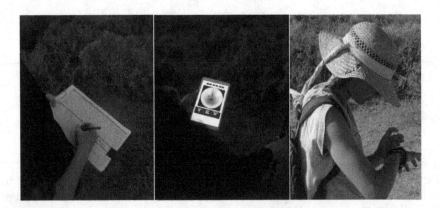

Fig. 7. From left to right: printed sheets used in traditional methodology, first version of LeafLab (Mobile), and the second version of LeafLab (Wear).

5.1 Hypothesis Test

Tables in [22] display the amount of time taken to complete each transect for the years 2016 and 2017, respectively. The classic statistic hypothesis t-student test (see [23]) was used to analyze the impact of these sample matches. Continuous variables are being used, and large samples have been obtained (n > 30), which adjust to a normal distribution as can be observed in Fig. 9.

This hypothesis test takes the following variables into account:

Variable x: Time required to finish a transect without context sensitivity.
Variable y: Time required to finish a transect using context sensitivity.

The null hypothesis and alternate hypothesis are as follows:

H0: There are no significant differences in the observed measurements.
HA: There are significant differences in the observed measurements.

The hypothesis test results are shown in Fig. 8a. The null hypothesis was disproved due to the drastically low p-value(less than a 5% significance level), and because the sample averages differ greatly (see Fig. 8b).

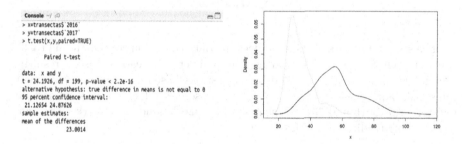

Fig. 8. From left to right: T-Test results and density plot for minutes in 2016 (black) and 2017 (gray)

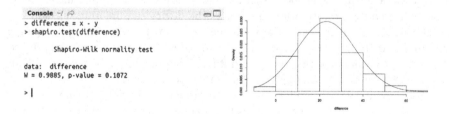

Fig. 9. From left to right: Shapiro-Wilk test and Difference Histogram

When analyzing the times registered for each transect completion, it was found that after implementing the context sensitivity upgrade at least 23 min on average was reduced from the overall time required for the survey tasks in each transect. The Shapiro-Wilk normality test when applied to the absolute differences (Fig. 9), yielded p-value>0.05. This allows for the difference variable (x-y) affirmation, which corresponds to a normal distribution displaying a 23.0014 min mean and a 13.44 standard deviation (Fig. 9 b).

The Cohen'D statistic (Fig. 10) for the estimation of the effect size revealed a large impact result (d > 0.8).

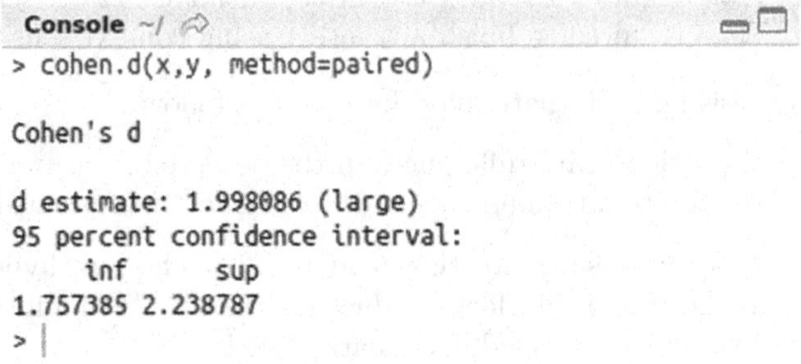

```
Console ~/
> cohen.d(x,y, method=paired)

Cohen's d

d estimate: 1.998086 (large)
95 percent confidence interval:
       inf       sup
1.757385 2.238787
>
```

Fig. 10. Cohen'D statistic

The statistic results obtained reflect the data collection corresponding to 400 transects, where at least 100 species were registered for each transect, resulting in a 40000 species registry, aided by suggestions based on context which made their task more efficient. The total for the absolute difference between samples registered for 2016 and 2017 measures up to 76.6 less hours of work when compared to the traditional methodology.

6   Conclusions and Future Work

Completing the preliminary groundwork, and obtaining real life results confirmed that the use of mobile applications and wearable technology vastly improve the user's experience and efficiency in these kind of tasks, when taking into account aspects like context information, as well as the device sensors and autonomy.

During this investigation, the tools and equipment used during campaigns were reduced to only two devices: one mobile device and one wearable device. The user must charge each battery and a few portable power banks (depending on the duration of the campaign) before setting out into the field.

In exchange, the user no longer needs to prepare and print datasheets or any other resources, folders, writing utensils, cameras, compasses, and gps devices. The development of this software truly innovated the productivity and workflow, because where they had previously only been able to complete around 27 transects per season they were able to increment this number to 200 transects with the application, which implies close to a 600% increase. Additionally, extracting the information is as simple as connecting the tablet to a wifi network, and they automatically have all their data available appropriately georeferenced, with the corresponding image relations and without any fear of input errors.

With this study we have been able to corroborate that when expanding the context sensibility in this way, not only does it provide a better experience for the user, but it also boosts the user productivity significantly.

Part of our investigation team is currently working on possible mobile applications in similar intensive data collection scenarios, where the manipulation of multiple sample objects is required for both field and lab work. A possible future goal would be to compare the data gathered in those new projects with the results exposed in this article.

Furthermore, even though all of the described software products are already in use, the possibility of including drones and image processing to improve the efficiency and labor conditions of these scientists has been contemplated for future proposals.

Acknowledgments. A special thanks to Emilio for his assistance and support, both during the experimentation periods and in the 2016 and 2017 campaigns.

References

1. George, D., Hatt, T.: Global Mobile Trends 2017 (2017). https://www.gsma intelligence.com/research/?file=3df1b7d57b1e63a0cbc3d585feb82dc2&download. Accessed 26 Mar 2019
2. Jarich, P., Hatt, T., George, D.: Global mobile trends: what's driving the mobile industry? (2018). https://www.gsmaintelligence.com/research/? file=8535289e1005eb248a54069d82ceb824&download. Accessed 26 Mar 2019
3. Mayuran, S., Pablo, I.: The mobile economy 2018 (2018). https://www.gsma intelligence.com/research/?file=061ad2d2417d6ed1ab002da0dbc9ce22&download. Accessed 26 Mar 2019
4. Whitworth, B., Ahmad, A., Soegaard, M., Dam, R.F.: The Encyclopedia of Human-Computer Interaction (2013)
5. Mostacedo, B., Fredericksen, T.: Manual de métodos básicos de muestreo y análisis en ecología vegetal. Proyecto de Manejo Froestal Sostenible (BOLFOR) (2000)
6. Spedding, C.R.W., Large, R.V.: A point-quadrat method for the description of pasture in terms of height and density. Grass Forage Sci. **12**(4), 229–234 (1957)
7. Elissalde, N., Escobar, J.M., Nakamatsu, V.: Inventario y evaluación de pastizales naturales de la zona árida y semiárida de la patagonia. Ed. PAN-SDSyPA-INTAGTZ, Trelew, Argentina (2002)
8. Fahrig, L.: Effects of habitat fragmentation on biodiversity. Annu. Rev. Ecol. Evol. Syst. **34**(1), 487–515 (2003)
9. Schilit, B., Theimer, M.M.: Disseminating active map information to mobile hosts. IEEE Netw. **8**(5), 22–32 (1994)
10. Abowd, G.D., Dey, A.K., Brown, P.J., Davies, N., Smith, M., Steggles, P.: Towards a better understanding of context and context-awareness. In: Gellersen, H.-W. (ed.) HUC 1999. LNCS, vol. 1707, pp. 304–307. Springer, Heidelberg (1999). https://doi.org/10.1007/3-540-48157-5_29
11. Almonacid, S., Navarro, P.: Aplicaciones móviles multiplataforma sensibles al contexto: una aplicación científica para el relevamiento florístico. In: XIX Concurso de Trabajos Estudiantiles (EST 2016)-JAIIO 45, Tres de Febrero, 2016 (2016)

12. Pazos, B.A., Morales, A.L.: Computación corporal: Expansión de la sensibilidad computacional hacia mejores experiencias de usuario. In: XXI Concurso de Trabajos Estudiantiles (EST)-JAIIO 47 (CABA 2018) (2018)
13. Wasserman, T.: Software engineering issues for mobile application development. In: FoSER 2010 (2010)
14. Charland, A., Leroux, B.: Mobile application development: web vs. native. Commun. ACM **54**(5), 49–53 (2011)
15. Que, P., Guo, X., Zhu, M.: A comprehensive comparison between hybrid and native app paradigms. In: 2016 8th International Conference on Computational Intelligence and Communication Networks (CICN), pp. 611–614 (2016)
16. Hassenzahl, M., Tractinsky, N.: User experience-a research agenda. Behav. Inf. Technol. **25**(2), 91–97 (2006)
17. Roto, V.: User experience from product creation perspective. Towards a UX Manifesto, 31 April 2007
18. Lu, Z., Chen, X., Li, Q., Zhang, X., Zhou, P.: A hand gesture recognition framework and wearable gesture-based interaction prototype for mobile devices. IEEE Trans. Hum.-Mach. Syst. **44**(2), 293–299 (2014)
19. Vanderdonckt, V., et al.: Distributed user interfaces: how to distribute user interface elements across users, platforms, and environments. In: Proceedings of XI Interacción, 20 September 2010
20. Demeure, A., Sottet, J.-S., Calvary, G., Coutaz, J., Ganneau, V., Vanderdonckt, J.: The 4C reference model for distributed user interfaces. In: Fourth International Conference on Autonomic and Autonomous Systems (ICAS 2008), pp. 61–69. IEEE (2008)
21. Elmqvist, N.: Distributed user interfaces: state of the art. In: Gallud, J., Tesoriero, R., Penichet, V. (eds.) Distributed User Interfaces. HCIS, pp. 1–12. Springer, London (2011). https://doi.org/10.1007/978-1-4471-2271-5_1
22. Firmenich, D.A.: Transects times (2019). https://figshare.com/articles/Transects_times/7862909. Accessed 26 Mar 2019
23. Rosenberg, J.: Statistical methods and measurement. In: Shull, F., Singer, J., Sjøberg, D.I.K. (eds.) Guide to Advanced Empirical Software Engineering, pp. 155–184. Springer, London (2008). https://doi.org/10.1007/978-1-84800-044-5_6

Author Index

Almonacid, Samuel   137

Barrientos, Ricardo J.   61
Basgall, María José   75

Cáseres, Germán   125
Chichizola, Franco   51
Corbalán, Leonardo   125

De Antueno, Joaquin   51
De Giusti, Armando   16
De Giusti, Laura   51
Delía, Lisandro   125

Estrebou, Cesar   51

Fernández, Alberto   75
Firmenich, Diego   137

Gómez, Pablo   110

Hasperué, Waldo   75
Hernández-García, Ruber   61
Herrera, Francisco   75

Igual, Francisco D.   3

Klagges, María R.   137

Lanciotti, Julieta   51
Lanzarini, Laura   98
Libutti, Leandro   51
Luque, Emilio   28, 61

Medel, Diego   110
Medina, Santiago   51

Morales, Leonardo   137
Murazzo, María   110

Naiouf, Marcelo   16, 75
Navarro, Pablo   137

Olcoz, Katzalin   3
Olivas, Jose A.   86
Ortega, Kevin   61

Paniego, Juan Manuel   51
Pazos, Bruno   137
Peralta, Daniel   61
Pesado, Patricia   125
Petrocelli, David   16
Pi Puig, Martín   51
Prada, Iván   3
Puig, Domenec   98
Puigbó, Alexandra Contreras   137

Quiroga, Facundo   98

Rexachs, Dolores   28
Rodríguez Eguren, Sebastián   51
Rodríguez, Christian   41
Rodríguez, Nelson   110
Romero, Francisco P.   86

Sánchez, Eduardo   86
Sosa, Juan Fernández   125

Tesone, Fernando   125
Thomas, Pablo   125
Tinetti, Fernando G.   41
Tirado, Felipe   28
Torrents-Barrena, Jordina   98

Wong, Alvaro   28

Printed in the United States
By Bookmasters